HISTORY OF THE *Metal-Casting Industry*

by BRUCE L. SIMPSON

second edition

AMERICAN FOUNDRYMEN'S SOCIETY, INC.

ISBN 0-87433-175-7

Printed in the United States of America

Re-Dedication

It is with very great pleasure that in celebration of the 100th Anniversary of the American Foundrymen's Society and during the term of my father, Henry Wallace Dienst, as the President of American Foundrymen's Society that our company, Simpson Technologies Corp., has the opportunity to join with American Foundrymen's Society to reprint this special edition of my Grandfather's book the "History of the Metal Casting Industry."

In the original "Salute to the Author" Mr. S. V. Wood, then President of AFA, said "I commend this volume to the reader, not alone for its historical value but for its stimulating recreation, its scientific facts and accurate data. It is an outstanding contribution, the result of patient, strenuous research born of an enthusiasm for an industry worthy of any man's energy and effort. It should be read and preserved by every foundryman."

My Grand father was truly a foundryman. In accepting the William H. McFadden Gold Medal in 1976 he expressed appreciation to the industry that had taught him a way of life. The privilege to be a part of the foundry industry has now passed to the fourth and fifth generations of my family and the feelings expressed so eloquently by my Grandfather, now more than 20 years ago, remain even more true for us today. For all foundrymen, and the foundry industry throughout the world, we are proud to offer you this special edition in celebration of our mutual history and one family's contribution to our extraordinary industry.

Bruce W. Dienst
President
Simpson Technologies Corporation
Aurora, Illinois
February, 1997

Preface

The author of this fascinating story is the President of National Engineering Company, today known as Simpson Technologies, and a past president of the American Foundrymen's Society. The first edition of this text bore an inscription by the 1946-47 Society president that bears repeating:

> This volume is but one of the great contributions which the Simpson family—Peter L., the father, Herbert S., the son and Bruce L., grandson—have made to the foundry industry. Pride of craft in each generation has been a constant inspiration to their associates.

History of the Metalcasting Industry is a tribute not only to this illustrious family but to all who have made metal casting a dominant force in worldwide manufacturing. The Society heartily recommends this unrivaled publication not only for its remarkable coverage but because it makes fast and highly interesting reading for the professional metal caster and his family and friends as well.

Contents

Introduction

THE METALCASTING INDUSTRY has a glamorous and thrilling past but one seldom used even by foundrymen in capturing the imagination of the public. It is hoped that this book will give foundries data upon which to build programs of public interest that will be informative to employee and customer alike.

This story of metal casting brings into sharper focus a fact confirmed by all scientific discovery, that progress is only as swift as research and equipment development permit.

It is an intention of this book to help bring about an increased interest in the unselfish promotion of better and more economical ways to cast metals and garner further support for those technical, trade and research societies that now are devoted to improvement of the science and technology of metal casting. It is hoped, too, that by showing the influence of accident and design on the casting of metals, it may be more widely recognized that all foundries have a common denominator regardless of the metal cast—that in any metal it is the fact that it is cast into shape that gives it real strength and reason for existence as units in a great worldwide industry.

No one could carry out the extensive investigational work of chronicling the developing of an entire industry without gaining a deep admiration for the men of the industry, upon whose willing and unselfish helpfulness a historian must depend for much of his information. This book could not have been written without the kindly cooperation and ready courtesies of many who, in the true sense, have made this project possible. Though it is impossible here to name all who contributed data, what follows is truly a foundryman's story, compiled by foundrymen and friends of the Industry.

The author is particularly grateful to those friends who loaned their private papers, sketches and photographs. Also of great aid were the files of technical societies, particularly the American Foundrymen's Society, and technical magazines and the trade press. To the many suffering librarians who endured with infinite patience and courtesy his frequent requests for foundry history materials, the author extends heartfelt thanks.

Perhaps no more fitting introduction to a history of metalcasting can be found than this quotation from the first foundryman to write at length on the subject, Vannoccio Biringuccio, head of the Papal foundry in Rome over five centuries ago–

> *I believe that my work would surely be almost a seed without fruit and that I would fail in that cause which disposed me to satisfy your request to write and form this work if, while laboring on it, I did not tell you of the art of casting, since it is a necessary means to very many ends. It is especially necessary since this art and work is not well known, so that no one can practice it who is not, so to speak, born to it, or who does not have much talent and good judgment. For this reason and also because it is closely related to sculpture, whose arms are the support of its life, it is very highly esteemed . . . it is a profitable and skillful art and in large part delightful.*

Bruce L. Simpson
Geneva, Illinois 1969

CHAPTER 1

Prehistoric Development of Metal

HISTORY is never complete. It is a constantly growing and developing thing.

As late as the seventeenth century, Archbishop Usher "determined officially" that human life commenced in 4004 B.C. This was accepted generally until less than one hundred years ago, when Boucher de Perthes of France, after years of ridicule, in 1849 won recognition for the claim that man was at least as old as the mammoths that once roamed Brittany and France. Since then archeologists, geologists and other men of science have steadily pushed back the curtain of the years with discoveries and studies that point rather conclusively to the existence of mankind at least 500,000 years ago.

The volume of new data constantly coming to light on our ancient ancestors is sufficient to cause any careful historian to mark limits of prehistoric fact with extreme caution. Constant excavations and research are continuing which, in another twenty-five years or so, well may render much of our present information obsolete or cast further doubt on what today may be generally regarded as fact.

In chronicling the development of metal casting, available data seem to indicate that the foundry had its beginning some four thousand years before the Christian era began. As a prologue to the advent of founding, a brief outline of man's history will enable us better to evaluate the extreme time required to reach the age of metals and (what proved to be of even more importance to civilization) the art of casting metal to form.

Historians and scientists have established divisions of prehistoric time in order to facilitate description, but for lack of exact data these divisions are essentially arbitrary. In many instances they overlap because progress follows no rule or pattern and may even develop at opposite ends of the earth simultaneously, thus making extremely difficult any precise and clear-cut chronological record.

Nearly a million years ago, at the opening of what is known as the Ice Age or the Pliocene epoch, it is believed that very human-like creatures lived on earth, fighting in packs for food and existence. This shadowy picture of man, or of his beginnings, still is cloudy a half million years later, at a time when the Java man, the Peking man, the Heidelberg man and other lower Stone Age creatures are said to have spread slowly throughout Asia, Europe and Africa, under semi-tropical conditions. Such men were little better than beasts, it appears, their weapons and tools probably nothing more than clubs and roughly shaped stones; and there is little or no evidence of their having even the first faint stirrings of that intelligence which finally set man apart from all other creatures.

About 250,000 years before Christ, the first significant developments of humanity occurred. The old Stone Age man made two important discoveries which definitely set him apart from the other beasts of the forest and field. Both discoveries undoubtedly were arrived at only after countless centuries of trial and error.

The first of these achievements was fire. Man's development of means for kindling and controlling fire for his own purposes evidences a tremendous intellectual advance.

It must be pointed out here that progress in those times was extremely halting and painfully slow. There was no flash of intuition such as marks the trained mind, but rather a long and arduous process of groping effort. Man undoubtedly was familiar with fire, from fires kindled by lightning or volcano; but the act of deliberately and successfully attempting to fire combustibles was by all odds the most remarkable and most important of all the acts of early man.

Simultaneously with fire, and perhaps even preceding it, came the discovery of flint and its use for tools. Prior to the use of fire and flint, man was a weak and puny being who lurked in the shadows and feared almost every other creature. At some stage in the dim past, however, some prehistoric man happened to pick up a piece of flint and found that it would shatter in such a way that a point or edge could be formed. How many thousand times man attempted (and failed) to shape rocks into weapons and tools cannot be recorded; but finally he was successful and adopted the process for his own use.

Logically, in chipping flint, sparks must have developed. Again for countless times this phenomenon was observed by man before the idea slowly sank in that the spark had a cause as well as an effect associated with the flint.

Possession of fire and flint enabled our prehistoric man to take his first long step forward. With these aids man lived in a cave, the door of which was guarded by a fire, and sallied forth in the daytime armed with pointed flint weapons. Such expeditions gave him both meat and clothing, which eventually brought about further developments.

In order to make clothing from skins, special awls and scrapers were developed, and his quest for skins further emphasized to man the necessity for specially shaped tools and implements. Fire and meat also were brought together, which called for containers... In turn, this meant fire-resistant (refractory) pots, a necessary prelude to metallurgy.

All of this slow and halting progress occurred generally some 125,000 years before Christ, and the slowly developing years brought forth a still more advanced type of human, the Neanderthal man. The climate of Europe and Asia again became cold, slowing further progress until the time of the Cro-Magnon man, the first true type of human, who appeared somewhere around 20,000 years before Christ.

Although thousands of centuries still were to pass before the dawn of civilization, in terms of what had gone before great progress was already being made. Men began to band together in villages and tribes for mutual protection and comfort. Animals were domesticated, and the first evidence of intelligent, planned behavior—agriculture, by systematic tilling of the soil, thus developing the first industry—is traceable to this period.

Early undecorated pottery now made its appearance, and many theories have been advanced concerning its origin. The most likely probability seems to be that pottery first was used in conjunction with cooking. For centuries man had eaten his meat raw; but sooner or later, probably by accident, he came upon cooked meat and liked it. Being man, he eventually got around to improvising ways and means for improving the task of cooking.

Perhaps the discovery of the refractory qualities of clay can be credited to a woman. In any event, evidence seems to indicate that meat was packed in leaves and covered with clay to prevent charring. When the combustible burned away, the fired shell remained.

For a long, long while primitive containers such as these usually were broken away from the daily meat and discarded, until a

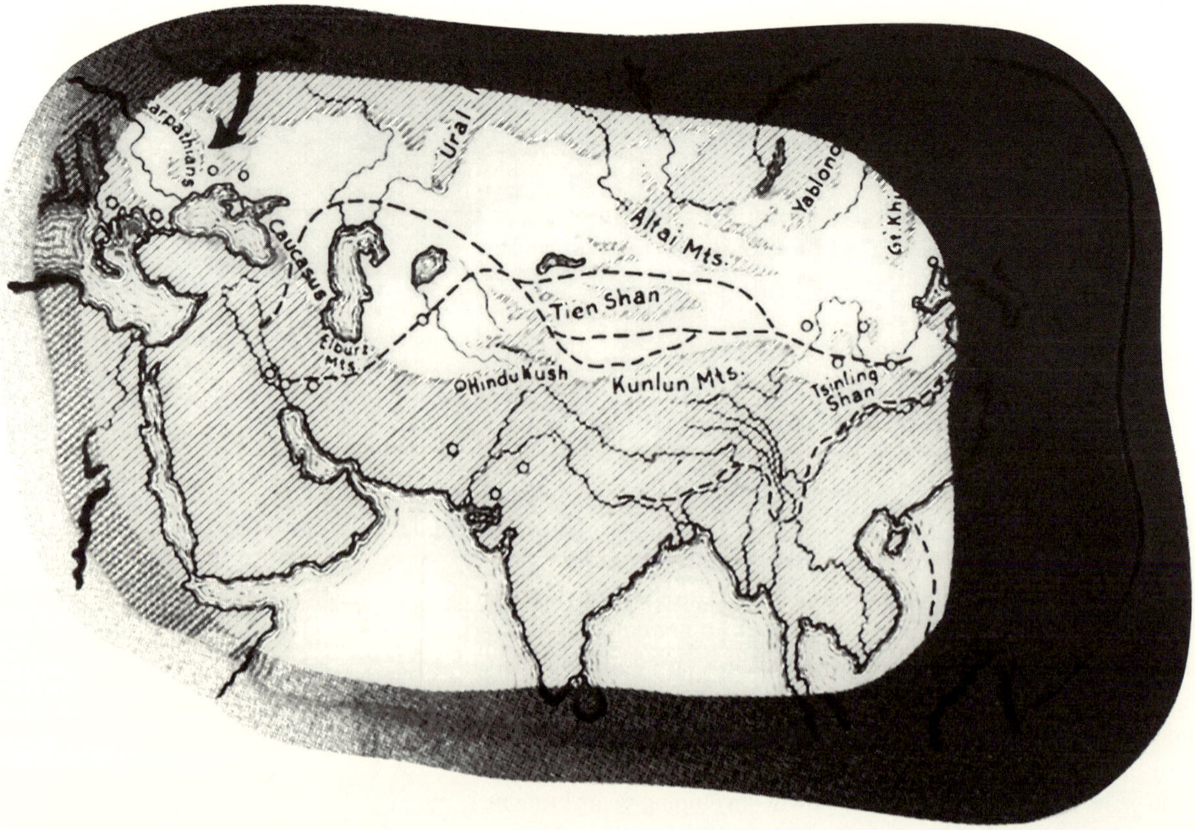

1 *The birthplace of metals has been traced with some accuracy to the steppe corridor of Eurasia . . . to the area north of the Black Sea in the Carpathian Mountains of Russia.*

combination of circumstances finally urged some prehistoric man or woman to examine the fired clay with other purposes in mind. From such curiosity developed stronger pottery for carrying water, grain and other objects.

Thus the stage was set for the coming of metals: fire to melt, and clay for crucibles to hold the molten product, were known. The utilization of native metals could now begin. The time was about 8,000 years before Christ.

The popular conception of prehistoric development is that the Stone Age extended down to about 3000 B.C., followed by the Bronze Age until 1200 B.C., at which approximate time the Age of Iron began. None of these periods, however, is sharply defined; in the first place, we find a period of gradual transition from stone to metal, then a change from pure copper to bronze.

The period of change between 8000 and 3000 B.C. is called the Mesolithic period of history and was marked by events that greatly stimulated human progress. Population moved in great migrations.

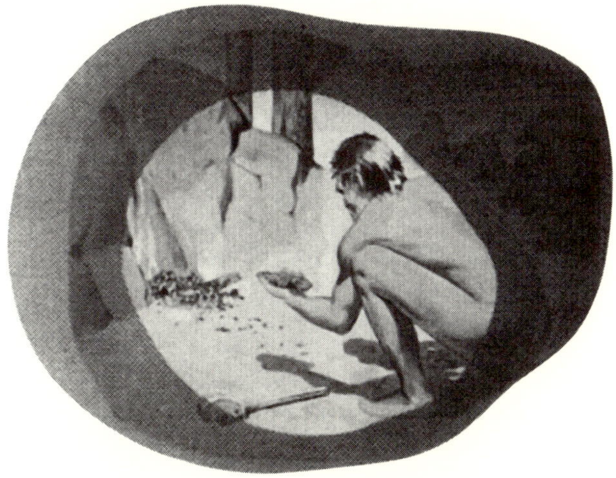

2 *How cast copper may have been discovered. It is believed probable that some native of the Near East may have banked his precious fire one night with copper-bearing ore, and discovered beads of mother copper on the hearth next morning.*

Wars among tribes and new tribal alliances encouraged discovery and accelerated development in many directions. Foremost among such developments was the use of metal.

There seems little question that the earliest notice of metal occurred in that section of the world known as the Eurasiatic "Steppe" belt, with greatest activity centered in what now is known as the Russian Black Sea area (Fig. 1). The people of that region were both industrious and warlike, a combination which historically has always spelled advancement.

It is accepted generally by scientists that gold was the first metal to gain attention, which is only natural. Primitive man had been picking up bright stones for centuries. The streams and waterways, his natural highways, of course held gold, and early man eventually picked up a pure nugget and admired its excellent surfaces. Soon after this, probably, came the realization that in this particular colored rock were qualities not heretofore noticed.

Primary in importance was gold's malleability, also its ability to resist corrosion and to remain gold without oxidation. Primitive man, having become accustomed to working with flint, probably expected the same type of fracture when gold was pounded; he was disappointed to find that the metal reacted quite differently. To his great surprise, the new material merely flattened, without splitting.

To man's primitive mind, this placed the metal in a classification of non-utility. Therefore, gold was generally relegated to the status of an ornament, although one great lesson had been learned: that gold was extremely malleable in the cold state.

The best of information indicates that gold was not melted or heated until after the advent of another new metal—copper. One might imagine that the discovery of copper came about through some process of reasoning, but the event seems to have occurred through pure accident.

It is fabled that a native banked his fire one night with copper-bearing ore and in the morning noticed some gleaming beads in the ashes (Fig. 2). This must have happened hundreds of thousands of times before some man, more observant than most, realized that copper was not gold and that the shiny beads had been melted out of the rusty looking ore which had been piled upon the fire.

At first these copper beads actually were confused with gold and, as with gold, were therefore used only for ornaments. As with gold, hammering of copper now was attempted; but here man became exposed to his first fundamental lesson in metallurgy. The copper became harder under hammering—and a new element for the prehistoric world was born. This was about 4,500 years before Christ.

For almost a thousand years the art of forging copper was developed in the making of weapons and implements. The process, however, had its limitations, as is well known today. Development of civilization brought about specialization and competition, both in peaceful and in warlike activities, and forged copper was limited in shape to a relatively few simple tools or weapons.

Man now was ready for, and prepared to make great use of, the cast metal shape. Fire was available, and refractory materials common and understood; the time had come for man to discover their vital relationship to molten metal.

Some four thousand years before Christ, the people of the Black Sea area swept down into Mesopotamia, conquering as they went, and their victories were due primarily to their forged copper weapons. Whether the conquerors or the conquered discovered the art of casting, is not now known; but in Mesopotamia, according to best information, a forge fire, operating under perfect conditions and reducing the copper ore to be hammered into tools, produced a castable molten product. The art of casting thus was born in the wondering eyes of some early foundryman who, entirely by accident, had combined the chemistry of metal with the mechanics of casting to shape.

3 *After the first open molds for casting copper were made in sand, man turned to "permanent" open molds of stone or sun-hardened clay. These open sandstone molds were used for casting (left to right) axe heads, spear heads, and lance heads.*

A whole new world now opened for the metal fabricator. Intricate shapes of all types could be made. The ladle began to supplant the hammer, and every primitive occupation quickly benefited. Horses could be harnessed properly for both war and peace. Agricultural implements appeared, shaped to accomplish specific tasks more quickly. Domestic wares and weapons became possible in shapes, sizes and types never before imagined. Men were quick to use their new-found knowledge, and the art of casting began to develop in earnest.

The first open molds were made in sand, but it soon was found that permanent open molds were more reliable. Molds then were cut in stone or formed in limestone or sun-baked clay, facilitating the production of more or less identical objects time after time (Fig. 3). A melting technique gradually developed, and a hollow log lined

4 *Open stone molds for the casting of implements and daggers, found at the site of ancient Troy. By that time melting technique had reached the stage where a hollow log, lined with refractory clay, had replaced a clay-lined hole in the ground as a melting unit.*

with refractory material was substituted for a hole in the ground lined with clay (Fig. 4). The melt was channeled into a reservoir, from which it was ladled into open molds by means of clay-washed containers.

Thus was foundry practice begun. The identity of the first caster of metal, and the exact spot where the first casting was made, are unknown and unimportant. The significant fact was the accomplishment. This was the beginning of the long process of development which, down through the centuries that followed, has resulted in the highly technical and specialized science of metal casting as we know it today.

The metal castings of ancient times were entirely pure copper, and were what now is know as "direct furnace pieces." Gradually the equipment of those early foundrymen underwent change, and the earliest change took place, quite naturally, in the melting medium.

At one stage of development it was noticed that the amount of metal produced was somewhat in proportion to the air draft, which gave a hotter fire. The first furnaces might have been built into the sides of hills so as to take advantage of natural winds, but man, as usual, found an easier and better way of doing his job. A new type of furnace was the result (Figs. 5, 6).

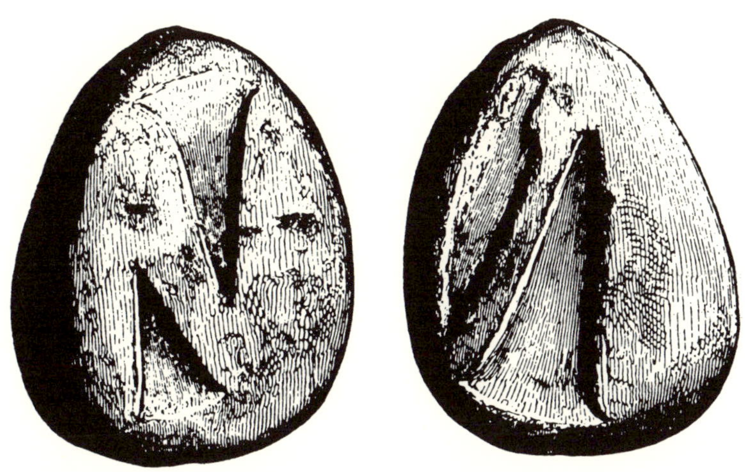

5 *Uni-valve stone molds dating back to about 3000* B.C. *Man by that date had conceived the idea of casting more than one implement at a time–in this case, three axes and a knife.*

6 *Stone and copper weapons of the Paleolithic Age. In pouring, metal was ladled out of a reservior by means of clay-washed containers.*

7 *Illustration of an early non-ferrous furnace. Ore was charged from the top of the furnace and blast applied through tuyeres of bamboo by means of foot-propelled bellows. Goatskin was the usual bellows material.*

Side elevation.

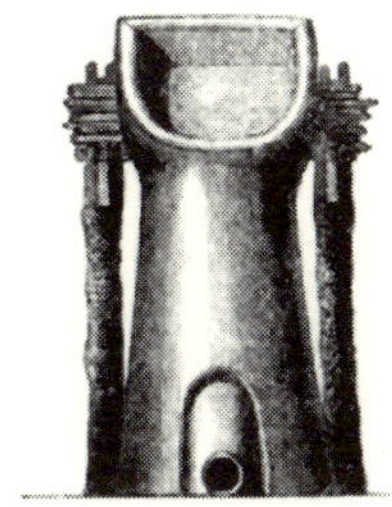

Front elevation.

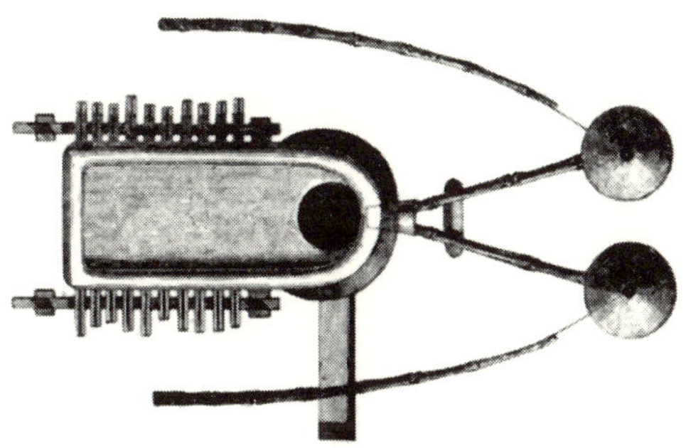

Plan.

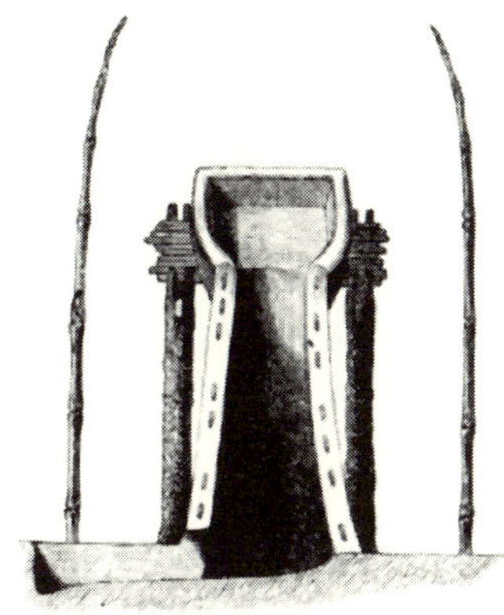

Vertical section

Vertical section of one of the bellows

Vertical section through the centre and one of the bellows.

Section of one of the bellows, showing the action of the heel in depressing the skin and closing the opening in its centre.

8 *One conception of how metal craftsmen of 3000 B.C. usually cast bronze tools and weapons in permanent stone molds. Even at that early date the usefulness of a bottom tap hole was known, as well as a clay plug or bot.*

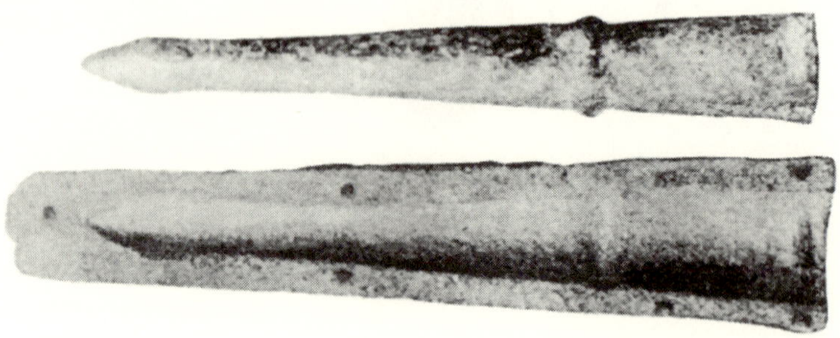

9 *Photograph of an early two-part stone mold. The lower photo reveals dowel holes for the cope section. Upper photo shows a plaster cast of a lance head. Foundrymen were beginning to think, to devise new methods, and to be more exacting in their work.*

10 *The oldest known casting in existence, a cast copper frog found in Mesopotamia, probably cast about 3200 B.C.*

An upright shell, lined with clay, was charged with alternate layers of copper ore and wood. Hollow wood blowers (tuyeres) were introduced into the furnace at the level of the firebox, with the first crude bellows at the blower end. These bellows consisted of a circular box covered with goatskin, with an escape hole in the center of the skin. The skin was tied to springy upright poles and the operator stepped on each bellows in turn, covering the escape hole with his heel and thus forcing air into the furnace. Release of the foot allowed air to flow back into the bellows, which then was snapped back to normal position by the poles. Alternate treading on the bellows delivered a fairly steady stream of air to the furnace (Fig. 7).

11 *Cast copper rein rings from Mesopotamia, dating back to 2700* B.C.

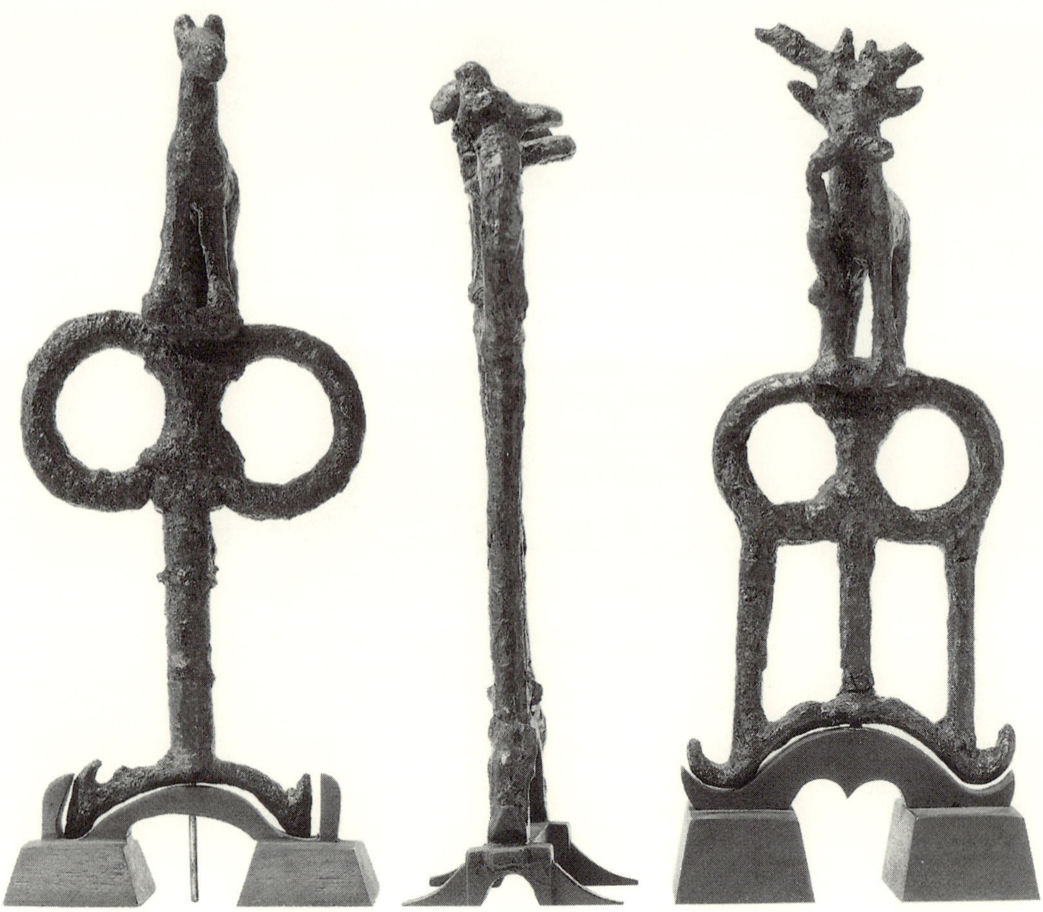

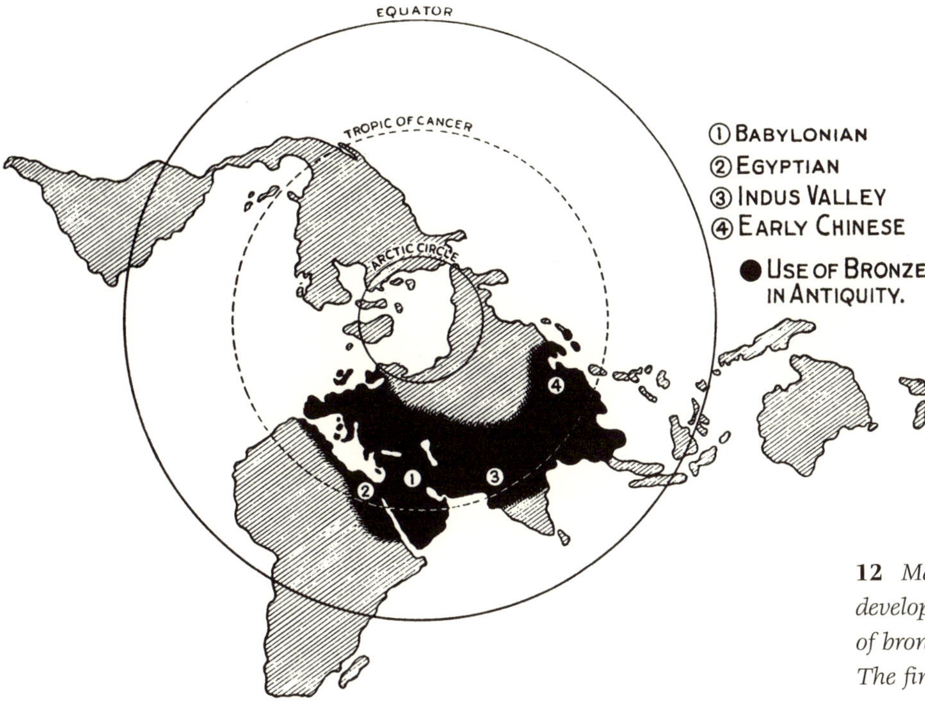

12 *Map depicting the probable development and spread of the use of bronze in the ancient world. The first accidental alloying of copper with tin in order to produce bronze took place several thousand years before the birth of Christ.*

The bottom of this "blast furnace" was opened by means of a taphole plugged by a clay retainer, which was removed when the operator was ready to pour.

Presently the desirability of casting to accurate shape assumed more importance, and attention next was directed to the molding process (Fig. 8). In open-face molding, only one side of the casting held a faithful reproduction of the mold. The top or cope section was missing, of course, which meant that the piece had to be pulled from the mold and the rough side hammered into an approximation of the desired shape.

To the more and more exacting foundryman this was a distinct handicap. To overcome it he constructed (much later) a complete mold in permanent fired clay, making it in two sections which he doweled together to insure a fit (Fig. 9). Into this completed mold he poured his direct copper; but, unfortunately, the foundryman soon encountered the first of his many troubles—surface blisters which completed ruined his castings.

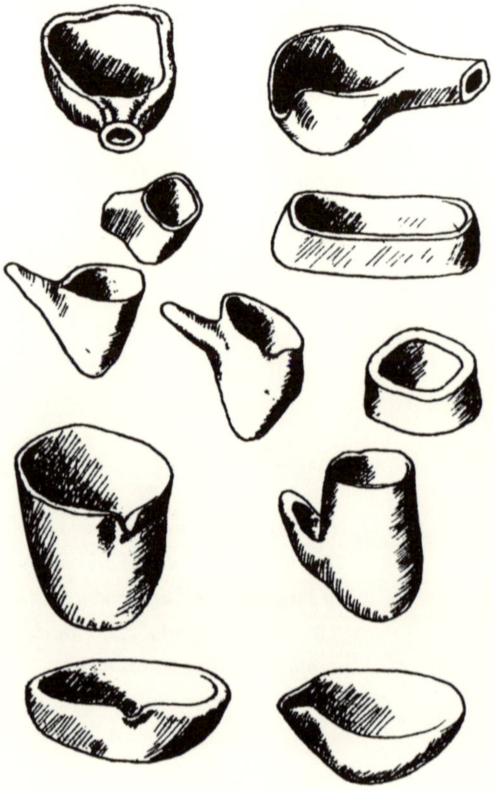

13 *Development of a crude "science" of melting, molding and alloying introduced the Bronze Age and the beginnings of modern civilization. Such crucibles as these were used during this period, commencing about 3000* B.C.

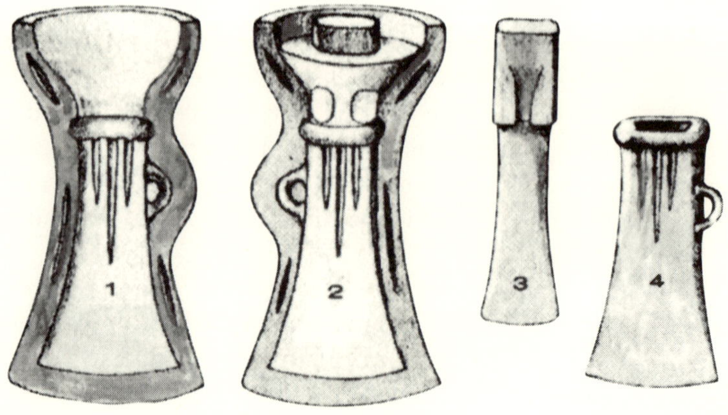

14 *Model of a bronze mold employed nearly 5000 years ago. Illustrated are (1) the empty mold, (2) complete casting in mold with core in place, (3) the core, (4) finished casting of socketed axe.*

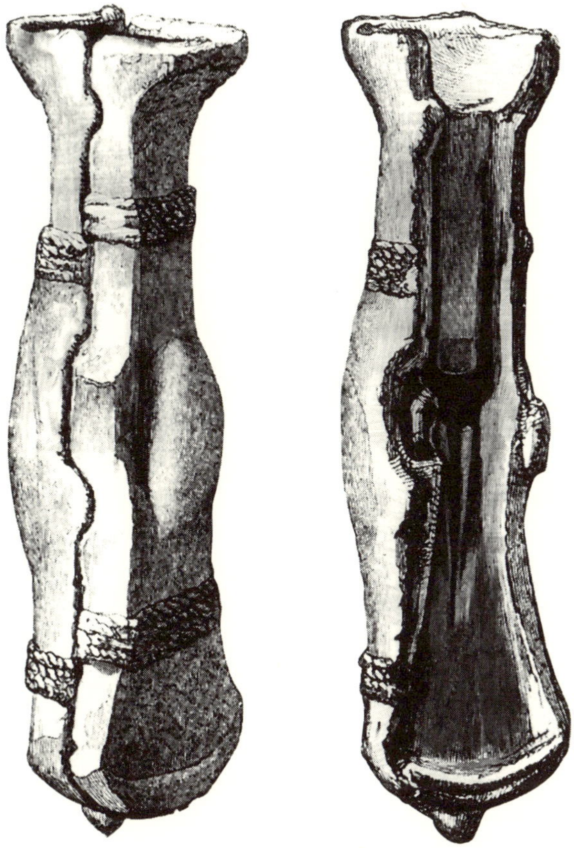

15 *Bronze mold for palstaves, found in England.*

The dream of a perfect, finished cast part thus seemed doomed, until accident again ultimately came to the rescue. For no apparent reason, some castings from complete molds were broken out perfectly, whereas others were not.

Finally the truth began to dawn on the early founder. He discovered that the good castings were made from ore containing something besides copper. Into some furnace at some time, someone had introduced an impurity that spelled the difference.

Actually, there was no immediate recognition of what was happening; as a matter of fact, the chemical answer to the problem was not known for another forty-five hundred years. It gradually became physically apparent, however, that a mixture (alloy) produced a better casting than did pure copper. The ingredient which accidentally had been combined with copper was, of course, tin. Thus the first alloy was born, and the chemistry of foundry practice entered the scene.

Not until the sixteenth century A.D. was tin recognized as a separate chemical element, but at this early juncture, by design or by accident, the foundry industry had taken a notable step forward. The science of melting, of molding and of alloying, crude as these techniques then were, had ushered in the beginnings of modern civilization—the Age of Bronze. The time was around 3000 B.C.

Several important metals by this time had become available. Silver, much more abundant than gold in a natural state but found less frequently in pure form, was discovered. Copper followed and, still later, the first use of iron.

In 2000 B.C., a date to which iron has been definitely traced, iron was regarded as a gem, good evidence of its novelty and scarcity. It is interesting to note, too, that almost every reference on iron points to the Bible statement that Tubal Cain was a "worker in iron."

It is definitely known that iron castings were made by the Chinese in the sixth century B.C. Since this is a story of castings, the exact chronology of metal is of only secondary importance. It is an historical fact, however, that no metal became important until man could cast it; and thus far only the beginnings of cast gold, silver, copper and the alloy bronze have been described. The foundry, the foundation upon which all modern engineering is built, nevertheless had already established itself among the arts and industries of early civilization.

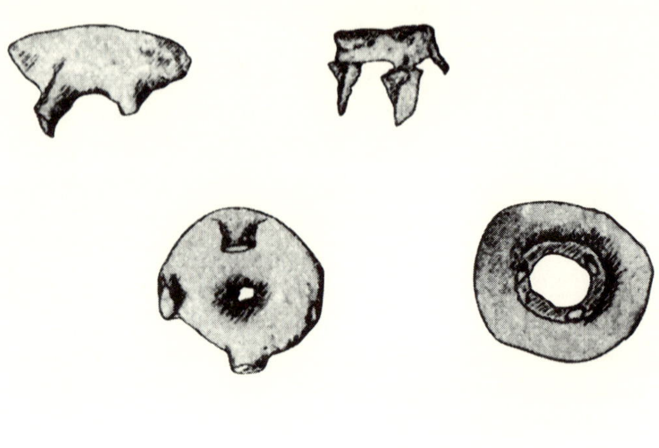

16 *"Gits" and runners, or sprues and risers, cut from ancient bronze castings.*

17 *An early ferrous furnace (about 2000 B.C.) for smelting iron. Fuel and ore were charged at the top, a lump or bloom of iron being pulled from the fire box and hammered into shape. This, of course, was not casting but forging.*

CHAPTER 2

Far Eastern Foundry Development

AT THIS point in tracing the development of cast metals, attention should be directed eastward into the Orient, for it was there that the art of casting metal first matured industrially, economically and scientifically. Early Oriental developments are important for two main reasons: First, the ancient foundrymen of the Far East achieved an exceptional mastery of bronze casting and, in a measure independently, an advanced knowledge of metallurgy.

The Orient also made two outstanding contributions to foundry practice, both of which appeared in the West much later as "new discoveries": in China they originally developed the casting of iron, and in India they invented cast crucible steel. These discoveries probably filtered westward into the European world, since the story of castings, both logically and chronologically, begins in Mesopotamia, moves eastward into the Orient, then turns back westward into the Near East, the Mediterranean basin, Europe and, finally, America.

Chronologically, foundry progress in the Orient practically ceased some time around 1000 A.D., while European achievements did not commence until about that time. Hence, before tracing the progress of metal casting westward from the Near East center of origin, it is best to record first the casting developments of China and India, during a period from, roughly, 2000 B.C. until 1000 A.D.

In speaking of Chinese history it is well to recall that the pottery or non-metallic era did not end until 2000 B.C. Less than five hundred years later we find a flourishing Bronze Age in full swing, with an extensive knowledge of the casting of metal (Fig. 18).

Research seems to indicate that China did not undergo any such long, painful period of growth as took place in the Near East.

18 *Hand operated bellows similar to those used in China for melting brass and iron in the sixteenth and seventeenth centuries B.C. The closed stone mold was the usual practice in China where the first cast iron was poured.*

This fact seemingly proves that information on metallurgy was brought into India and China from Mesopotamia, probably by caravan or by some wandering tribe. It apparently was passed on verbally, since no written record of transmission has ever been discovered. Even at that early date, foundrymen were rather inarticulate as to their accomplishments, but, fortunately, the product they made has survived the years.

It is generally accepted that bronze founding appeared in China as a fully developed technique, as an introduction from western countries. China reveals no copper-working prologue to the flourishing bronze industry of the Shang dynasty (1766-1122 B.C.). While Chinese annals claim that founding was practiced in earlier times, from about 2200 B.C., archeologists have unearthed in China no cast metals predating the Shang bronzes, regarded as the most perfect of their kind.

The "Great City Shang," Anyang in the Hwang valley, seems to have been China's first important foundry center. Its chief cast-metal products included a great variety of complicated and fantastically decorated ritual vessels, dagger-axes, adzes, knives, arrowheads, spears and the like. The Shang motifs are said to have been elaborated from earlier models in pottery and wood, and became permanent elements of design.

China's expert bronze founders, according to Confucian scripture, were required to give "the suitability of the article the first consideration," and were held guilty when their workmanship was not "what it ought to be." During the Chou dynasty, succeeding the Shang period, many ritual bells and musical rattles were cast, as well as those picturesque Chinese coins which were miniatures of the knives, sickles and other implements previously used as articles of exchange. When the aggressive feudatory nobles became China's real rulers, more attention was given the casting of weapons, shields, bits and bridles, and chariot fittings.

Three dynasties or periods mark the era before Christ in China. First was the Shang dynasty, which lasted from 1766 to 1122 B.C; then the Chou dynasty, between 1122 and 255 B.C; and finally the Han dynasty, from 206 B.C. to 220 A.D.

The earliest known Chinese cast statues date from the short Ch'in dynasty (249-206 B.C.). Chinese foundrymen were considerably experienced in casting colossal images. Outstanding among the work of the Han founders were their cast bronze mirrors.

During these periods Chinese metallurgy was greatly advanced. Little or no copper forging was practiced, but widespread use was made of the closed stone mold and the foot-bellows furnace previously described in connection with prehistoric foundry practice.

19 *Copper token money used in China was cast in the shape of swords, hoes and other common objects. While the Chinese excelled in the chemistry of metals, few pure copper objects have been found there.*

The Chinese particularly excelled in the chemistry of metals. They experimented with and developed the principles of metal mixtures for different objects, dependent to a great extent on how the objects were to be used. Here are a few early mixtures employed:

Bells and kettles	5 parts copper to 1 part tin.
Axes	4 parts copper to 1 part tin.
Knives	3 parts copper to 1 part tin.
Spades and hoes . . .	2 parts copper to 1 part tin.
Mirrors	1 part copper to 1 part tin.

Thus, for the first time in history, foundrymen looked to the use of their products—to their "suitability"—as a guide for their workmanship.

China was fortunate in having mineral deposits which contained large supplies of both copper and tin in close proximity. Both of these essential elements for the making of bronze were particularly abundant in southwestern China, and the foundrymen made liberal use of them. At the same time, metals and ores may have been drawn from many areas, since copper and tin seem to have been demanded as tribute materials during the Chou dynasty.

Chinese bronzes of this period have been coming to light in recent years, and the United States is fortunate in having one of the finest collections in the world. The Freer Gallery of the Smithsonian Institution in Washington, D.C., has a remarkable collection, and the illustrations of early bronze castings shown in Figs. 20-23 are taken from this source. Most of this work was made by the "lost-wax" method of molding, which will be described later, and it must be admitted that the work is excellent. Extreme patience in both molding and melting is exhibited in these masterpieces.

20 *A beautiful cast bronze box dating from China's Shang dynasty (1766 to 1122 B.C.). The delicate filigree was achieved by the lost-wax process of casting, still used today in casting many of the finest art bronzes.*

21 *Bronze libation cup which, like the bronze of Fig. 20, was cast during the Shang dynasty.*

22 *Bronze ceremonial elephant cast in two parts and joined, depicting the workmanship and patience of the early Chinese foundrymen. From the Chou dynasty (1122 to 255 B.C.) or earlier.*

23 *A cast bronze blade intended for a shafted weapon, probably a work of China"s Chou dynasty.*

Copper ore, charcoal and slag, pottery crucibles and many mold fragments have been found on the site of the "Great City Shang." Piece molds, keyed with tenon and mortise, hint at the methods used. Lost-wax techniques, closed stone molds and sectional loam molds are indicated. Although both the Shang and Chou artisans turned out some intricate one-piece castings, separately cast and assembled parts seem to designate the more usual practice.

One interesting sidelight on molding during the Chou dynasty should be noted. Particularly noticeable in Fig. 24 are the projecting flanges on corners of square bronzes, or dividing the surfaces of round bronzes. These flanges apparently had their origin in the projecting fins left on castings made in two-piece molds. Upon the advent of one-piece molds, these fins were deliberately copied and molded, probably because they were observed on the originals. In other words, the molder apparently felt that in some way the fins had contributed to the making of a good casting; so he deliberately continued them for no apparent reason, until they gradually fell into the class of ornamentation.

24 *Cast bronze bowl of the Chou dynasty. The projecting flanges apparently were copied without question from the metal fins which previously remained at the time when castings were made in piece molds.*

One cannot help but be impressed by the fine work accomplished by those early Chinese founders, proving their extraordinary ability not only artistically but also from a metallurgical point of view. Most of the objects found show little chiseling, filing or cleaning—a tribute to foundry ability.

In the casting of iron, Chinese history again does not follow the usual pattern of trial and error. While the Chinese were not original in much of their progress, they seem to have excelled in developing the techniques delivered over to them. Undoubtedly word had reached China from the outside that a new metal was available with properties in some ways superior to bronze. The Chinese seized this opportunity and immediately developed cast iron without waiting for the long process of gradual evolution through forging and blast furnace. Considering their tools and the fact that the western world did not achieve this growth until nearly sixteen centuries later, it is almost unbelievable; but the fact remains that cast iron was produced somewhere around 800-700 B.C., and the earliest known sand mold has been traced to 645 B.C.

This extremely rapid progress has amazed metallurgists in recent investigations, and it is interesting to note how they account for the sudden appearance of iron in cast form. The major reason given concerns the fact that the Chinese had developed melting equipment capable of producing greater air draft than hitherto had been possible. Methods are set down by one Meng Hsi Pi Tan which describe the box-bellows furnace (still used in China), the most advanced melting unit of the time. A modern translation of his description of the melting of both iron and steel is worth of reproduction here:

"Iron ores are smelted with charcoal in a shaft furnace. The blast is obtained from a box-blowing engine of wooden construction. A piston is moved backward or forward by human or animal power. The air enters through openings provided with valves in the side of the box, and passes out to the furnace through a passage formed by placing in the box a perpendicular parting wall connected with the piston chamber by valves at both ends. From this passage the air passes though a tuyere into the furnace (Fig. 25).

25 *Drawing of a Chinese cupola which, while modern in some respects, depicts the improved bellows box. Illustration shows: (a) top section, which could be lifted off; (b) tuyeres; (c) box type bellows; (d) double bellows piston; (e) bellows drive handle; (f) lower section of cupola; (g) pouring spout feeding into metal reservoir.*

"From the furnace the metal produced flows to a reservoir and from there it is ladled out into molds for castings. The furnace for the manufacture of iron is made of clay mixed with salt. A hollow in a mountain is sometimes lined with clay, or a large wooden mold is used and the furnace is allowed to dry for a considerable length of time. Such a furnace holds two or three tons of ore."

Metallurgists advance a second valid reason for the abrupt shift to iron in China, involving procedures which the Chinese evidently understood but which the European world did not employ for many centuries. It has been determined that oxide of iron, when heated at a low temperature in the presence of excess carbon, is reduced to a soft, pure iron which cannot be melted in small furnaces, since the melting point of this soft mass is some 1530° Cent. (2786° Fahr.). However, if the iron is held at a high heat while in contact with carbon, it absorbs carbon, and the melting point drops to 1170° Cent. (2138° Fahr.). Chinese foundrymen, with their advanced design of draft and construction of melting furnaces, understood this procedure and thus obtained liquid iron long before it was produced regularly in other sections of the world.

26 *Front view of oldest known Cast iron stove of the Han dynasty (206 B.C. to 22) A.D.), the last of China's great dynasties. Inscription reads "May it please your lords."*

27 *Side view of same cast iron stove illustrated in Fig. 26, showing the integrally cast legs.*

28 *Cast iron ink pallet attributed to the Han dynasty, used for the mixing of inks. Note the delicate base, molded by use of paper patterns.*

In order to reduce the pouring temperature, the Chinese made many additions to their melt which enabled them to cast the fine, thin-section work that they have left as models of early foundry practice. One such addition was knows as "hei-tu." When analyzed, this material proved to be a high-phosphorus coal.

In addition, the iron ore used was high in phosphorous content, which produced a six or seven percent phosphorous iron, making the metal more fluid and allowing it to be poured at around 980° Cent. (1796° Fahr.), or 100° Cent. (212° Fahr.) below the melting point of copper. These advantages, plus the presence of abundant wood for charcoal and ore that could be obtained from open pits, made it possible to produce an economical and excellent cast-iron product.

These advantages were further increased by the discovery of coal and iron ore in close proximity. When coal finally was used as fuel, it was found to be very low in sulphur. Thus nature aided the Chinese in their quick perception of the value of cast iron.

Early evidence of iron castings in China has been explored by top-ranking American archeological and metallurgical experts,

29 *Cast iron vase of the Han dynasty, with cast iron handles. It is believed that the handles were cast first then filmed with sand and laid into the main mold, to be joined to the vase as the latter was poured.*

among them Berthold Laufer of the Chicago Museum of Natural History and Professor Thomas T. Read of Columbia University, New York. Prior to their discoveries and analyses, we were forced to depend on Chinese written records alone, which nevertheless have been found to be extremely accurate. Earliest authenticated evidence is a record dating back to 513 B.C. of 600 pounds of iron being requisitioned to cast a tripod on which the criminal code was inscribed. Cast-iron plowshares were recorded in 233 B.C. The Laufer castings illustrated in Figs. 26-31 are believed to have been made in the Han dynasty around 200 B.C. and are the oldest castings of iron known to exist.

It is interesting to note that iron founding was a state monopoly at that early time, and a special bureau in China was established for its control. Anyone engaging in the manufacture of iron castings was severely punished if found not to be properly accredited. The castings of Figs. 26-31 illustrate the earliest cast-iron stove, and vases and fittings of all types.

30 *Cast iron pan of Han dynasty, one of oldest iron castings known to exist.*

31 *Cast iron fittings, again of the prolific Chinese Han dynasty. Left to right: bell, axe head and chariot hub cap.*

Professor Read has analyzed, both physically and microscopically, many other early iron castings. From his studies we have learned that the carbon content of these pieces ranges from 3.1 to 4.5 percent, and that the microscopic analysis shows a sound metal structure.

Among these pieces is a recumbent lion (Fig. 32), the earliest dated iron casting found in China, cast in 502 A.D. The statue of Kwan Yin (Fig. 33), the Chinese goddess of mercy, was cast in 558 A.D., and the large lion now standing in a monastery yard near Ts'ang-Chou (Fig. 34) was cast about 953 A.D.

There is a fable to the effect that a local foundryman was found guilty of a criminal offense and cast the Ts'ang Chou lion to guard the city, thus gaining pardon for his crime. It was cleverly cast by pouring successive layers, as with a monolithic concrete building. The casting is nearly 16 feet long and about 20 feet high and was a truly remarkable accomplishment.

32 *Cast recumbent lion, the earliest dated iron casting (502 B.C.) found in China. From the private collection of Prof. Thomas T. Read of Columbia University, New York.*

Cast iron served the Chinese well and finally became so common and useful that it was used for the roofs of pagodas built around 1100 A.D.—and still in excellent condition (Fig. 35).

Just across the China Sea lay Japan, where the foundry art soon was introduced, undoubtedly wholly influenced by Chinese developments, including the box-bellows furnace. Early Japanese bronze castings—swords, spears, arrowheads and tools—were made in stone molds, but lost-wax molding soon became the dominant technique (Fig. 36). Loam and sand molds, and iron working as well, may have been carried to Japan by Chinese founders in the second century B.C. However, few records have been found of metallurgical history in Japan.

33 *Cast iron statues of Kwan Yin, Chinese goddess of mercy, cast about 558* A.D.

34 *This large cast iron lion, cast in China about 953 A.D., still stands in a monastery yard near Ts'ang-Chou, Hopei province. The lion measures 16 feet in length and 20 feet in height, and was cast in small sections. Metal thickness, 2 to 8 inches.*

The brilliant age of Japanese bronze founding dates from the introduction of Buddhism, in the sixth century A.D. to the thirteenth century. As is the case with all new religious movements, there followed a great deal of scrapping of old shrines and the erecting and outfitting of new ones. Among their creations were the masterful trinity groups; the colossal seated Buddha of Todaiji (about 54 feet high); and the great temple bell in Nara, with a diameter of 9 feet, height of 13 feet and metal thickness of 8 to 10 inches.

The Japanese endeavored to build each new shrine to Buddha larger and larger, until finally, at Kamakura near Tokyo, the Great Buddha was cast in 1252 A.D. "Diabutu," as it was called, was cast in bronze and weighed 92 tons, with a height of 42 feet and a 97-foot circumference. Such bronze colossi undoubtedly were cast in sections and, since no seams are visible, may have been "burned together" or "united by running in an alloy containing large proportions of tin and lead." The iron founders of Japan seem to have achieved considerable skill under Chinese instruction, and also to have gained high social position.

35 *Cast iron was so widely used in China around 1100 A.D. that it was employed to roof pagodas and other structures. This Chinese temple roof of cast iron still is in excellent condition. Incense burner in front of the temple also was cast during the eleventh century.*

Aside from such religious works, foundry activities in Japan added little to the foundry art. Most of the technicians seem to have been Chinese, and the methods employed thus were close copies of those already prevalent in China. There is no evidence of any new or unusual developments in the casting of metals in early Japan, all efforts seeming to have been directed toward the making of larger and larger statues.

This was not the case in India, which had a long history of metal-working. India's metallurgical efforts evidently began early and were actively pursued, but the main drawback to investigation today is the lack of definite information and data. This casting of bronze, although it apparently was practiced at as early a date in India as in China, seems not to have achieved the prominence or the artistry of the Chinese work.

Some excavation in the Indus valley has revealed highly cultured cities, founded perhaps about 3000 B.C. by people who knew copper and cast tools and weapons in closed molds and, working later with bronze, turned out especially fine figurines by the lost-wax process. The metals in use–copper, tin, gold, silver and lead—were smelted in brick furnaces, patterned after pottery furnaces, and probably were remelted in crucibles for casting.

Some records, dating back to around 1000 B.C., refer to metal-workers and iron, but the Hindu never really discovered cast iron. The early ironworkers were probably smiths, not foundrymen. While they may have obtained liquid iron at times and realized that it was castable, there is no clear evidence of an iron castings industry in India prior to about 200 A.D.

The Hindus, fully as intelligent as the Chinese, did unusual things with metal but lacked the persistence of the Chinese. Furthermore, their accomplishments are shrouded in the mysteries of the cults and sects which controlled the shaping and casting of metal and greatly influenced the growth of industry. Still, the Indians unquestionably were the developers of cast steel, which is most surprising even though the availability of minerals undoubtedly was a contributing factor.

India early developed the wrought iron or "blooms" and removed impurities in the metal by forging. Examples of forged work date back to a thousand years before Christ, or even earlier, and include such famous works as the Pillar of Delhi. Records are exceedingly sketchy, but probably after the casting of iron in China the Hindus developed the crucible process of casting steel. Centuries later this was credited to Huntsman in England, but whether or not this was an independent discovery may never be known.

36 *From China the foundry art spread to Japan around the time of Christ but few Japanese records are known. The Philippine forge furnace illustrated is believed similar to those used in Japan and the Far East and employs the piston type of bellows blast.*

37 *Cast steel unquestionably was first produced in India, where crucible steel has been traced back to 500 A.D. The illustration depicts a method of casting crucible steel similar to the methods used in India. There "wootz" steel was produced by placing wrought bars in crucibles and melting them under a long and intense heat.*

This much is certain, that the Indians knew the principles of cast crucible steel somewhere around 500 A.D. They had taken wrought bars and annealed them in a charcoal fire until a type of steel was obtained known later as blister steel, or cementation steel. Not content with this, they put the bars into clay crucibles (of unusual strength, probably due to local conditions) and by means of a long and intense heat obtained a liquid steel known as "wootz" steel.

No further details of this crucible process have ever become known, except that the Hindus claimed great benefits from wrapping the bars in certain green leaves before they were placed in the crucible. In any event, this process fell into discard and was lost until the eighteenth century A.D., when Benjamin Huntsman finally perfected a crucible that would stand the heat necessary to produce liquid steel. India's ironworkers obviously had been on the verge of a startling discovery. Had they gone a step further, the development of civilization might have followed entirely different channels.

38 *An open sandstone mold for casting copper and bronze daggers, chisels and ingots, of a type used in Egypt around 2500 B.C. Gypsy metallurgists are believed to have carried the knowledge of metal casting from Babylonia and Mesopotamia into Egypt, Palestine and Asia Minor, about 2800 B.C.*

39 *Castings made from the same mold shown in Fig. 38. The upper casting is a copper chisel, taken from the vertical section of the mold, at right. From the horizontal section at the left came the dagger shown (top).*

Here the history of Far Eastern foundry development ends, for after the tenth century A.D. the Orient closed in on itself, and metal-working progress ceased. Because there was little opportunity for transmittal of information, particularly that of a technical nature such as can be obtained through our foundry associations and patent system of today, most of the work of China and India had to be painfully and patiently repeated. The impetus of discovery and exploitation continued in the Mediterranean basin and Europe, where men were actively using and advancing their metallurgical knowledge.

CHAPTER 3

Casting Development in the Mediterranean Basin

IN TRACING the development of metal, and particularly of castings, the trail begins first in Mesopotamia, then splits east into China and India, and finally turns back west into Egypt, Greece and Rome. Thus it has been necessary to consider Chinese and Indian foundry progress before following the growth of civilization and the casting of metal in the Near East.

By approximately 2800 B.C., we may be certain, the metal-casting achievements of Mesopotamia and Iran had become known in Egypt, Palestine and Asia Minor. Itinerant foundrymen may have carried first the finished product and then the method of processing into these lands. Migrations and invasions, the growth of trade and commerce, and quests for ores and other raw materials also account for the diffusion.

The arts of smelting copper, alloying with tin into bronze, and casting it first into open and finally into closed stone molds, appear to have been introduced to the early Egyptians as well developed techniques. The Egyptians in the first stages of their remarkable growth were an intelligent and resourceful people, and it was not long until they began to make technical improvements in the casting art. Powerful and rich, they early appreciated both the useful and the artistic value of metal, and presently made two of the most important advancements in metal production.

One concerned refinements in the smelting and forging of iron, which will be discussed later; the other was the discovery of a molding method now known as "lost-wax molding." This investment or destructible pattern molding has retained its importance for more than thirty-five hundred years, and still is used today in

40 *Shaping a pottery vase in early Egypt. Cope, drag and core molding seem to have appeared early in Egyptian history, combining a knowledge of both pottery and metallurgy. Molding developed from pottery making, particularly molding for the lost-wax process.*

the making of dentures and art bronzes. The process has also come into favor in the casting of ferrous alloys and is one of the factors that made possible the superior performance of our jet-propelled aircraft. Without doubt this method of molding, to which the world has given little acknowledgment, ranks as a top-flight achievement without which metal could never have gained its early prominence, and the loss of which undoubtedly would have retarded the growth of civilization in large degree.

"Lost-wax molding" has been a much abused and misunderstood term. Some otherwise reputable researchers have stated seriously that it is a "lost" process, discovered in Egyptian times and then lost until the fifteenth century A.D., when it was presumed to have been rediscovered. This thesis completely ignores the thousand upon thousands of castings made in the interim—castings which could only have been cast by the lost-wax process—and completely misinterprets the significance of the word "lost" (Fig. 40).

Cope, drag and core molding apparently appeared early in Egyptian history, and without these developments the foundry industry still would be insignificant instead of the giant it is today. The unknown man or men who developed these methods brought together, in happy combination, the knowledge and experience of two important industries, pottery and metallurgy. Thus began the close association between pottery and molding which was maintained up to as recently as two centuries ago, when the average molder apprentice was drawn from the ranks of potters, and the iron-foundry molder actually was known as "the potter."

The close ties between molding and ceramics seem to indicate that they developed almost simultaneously. It was the potter's art—the selection and compounding of suitable clays, their molding and proper firing–that gave the foundry the crucible. The same designs in pottery and in the castings of early years point with certainty to the fact that the same man made a ceramic piece and also the clay and wax mold into which its metal counterpart was cast.

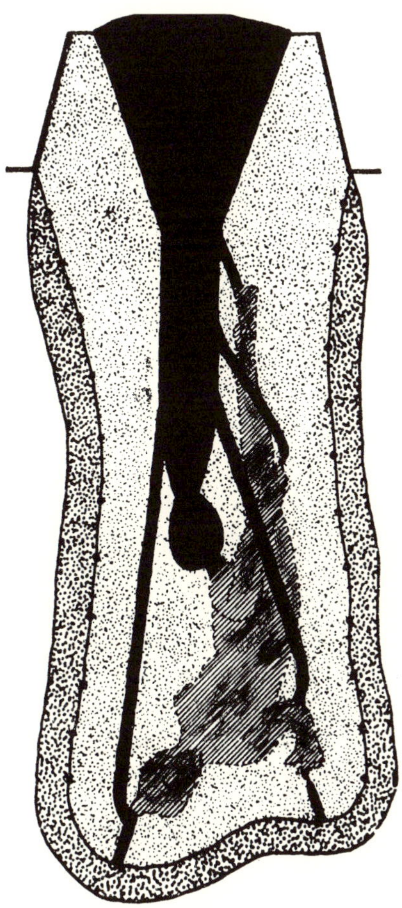

41 *Gated and invested model of statuette, showing feeds and cavity left by the melted wax in the lost-wax process. This piece contains no core.*

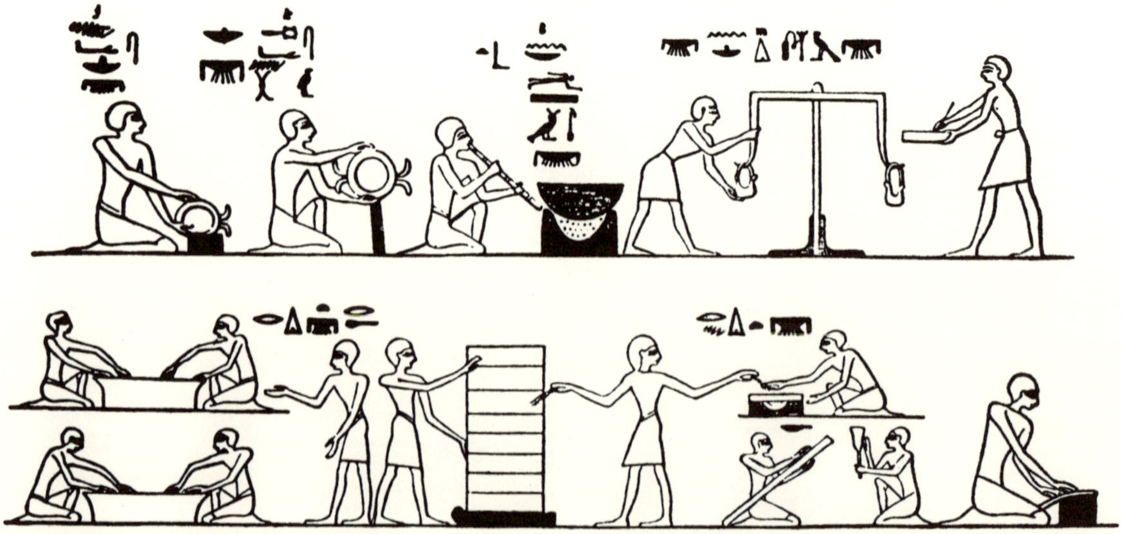

42 *Egyptian goldsmiths preparing molds, melting, shaking out and weighing cast gold jewelry. Note the use of blowpipe and crucible for melting.*

Potters developed the first molding tool, the potter's wheel, and it was left to some unknown genius to whom the world still is indebted—perhaps a potter-molder, a craftsman in the workshops of one of the early Pharaohs or high priests—to invent lost-wax molding and, simultaneously, transform a pottery piece into a metallurgical tool.

The idea probably originated as the potter, working at his wheel, created a ceramic piece. While etching on the design, prior to firing, it occurred to him that the piece would be more attractive and more stable if made in metal.

Shaving off the etched design, he first reduced the outer dimensions of his piece to a proper thickness and then coated the piece with sufficient wax to regain actual size. In the wax he redrew his designs or shapes. Upon this wax-coated mold he then carefully plastered layers of clay and bound the entire mass together, after filling the cavity of the piece proper with rubble and scrap clay. When this was completed, he reversed the mold and baked it in the kiln in a manner quite similar to the firing process employed in ceramic work.

Under heat the wax ran out and, although termed "lost," actually was collected for re-use. Finally, into the completed mold he poured an alloy of copper and tin, and when the metal had cooled the mold was broken away, revealing in metal a completely cast piece.

Considering this early artisan's background of experience, his was an innovation of the greatest importance. No other process can be named by archeologists so crowded with deduction, engineering ability and ingenuity. When one considers the meager tools, education and general experience of the average man of those early

43 *Egyptian painting depicting ancient foundrymen melting gold. Note foot bellows, pouring the gold (ladle handled by means of sticks), blowpipe and crucible.*

44 *Egyptian coppersmiths melting with foot-operated bellows draft, a method probably introduced from Asia Minor. From a tomb painting about 1471* B.C.

times, lost-wax molding must be classed as an achievement equal to the invention of the wheel or the discovery of fire. In one operation, man had developed the idea of a matrix, which he was also clever enough to remove without destruction or distortion of the mold itself.

With the wedding of the metallurgist and the ceramic craftsman, the foundry art now could advance with astounding speed. From this point forward, only the limitations of equipment and failure to appreciate basic principles delayed the foundryman in achieving his remarkable destiny.

The spread of metals and founding traveled from Egypt, Syria and Asia Minor, and from Mesopotamia, throughout the length and breadth of the western world, and the search for new metal sources was almost entirely responsible for the expansion of civilization

45 *Bronze foot cast in Egypt. Note the sprue still attached to the heel.*

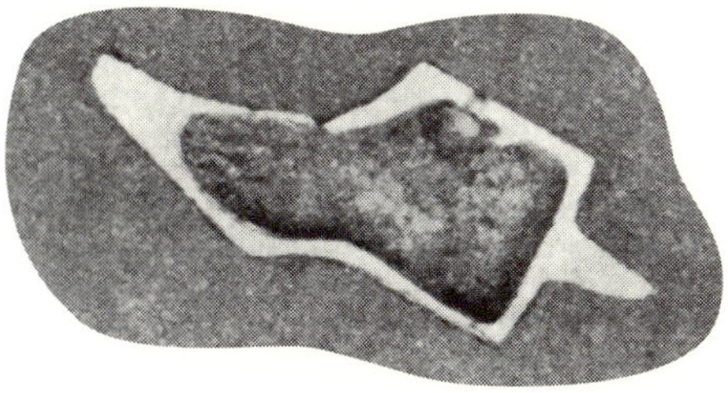

46 *Section of Egyptian bronze foot illustrated in Fig. 45, showing that the core was not removed.*

47 *Unfinished casting of a bird's feet, from Egypt. Note design of runners and gates which fed the casting.*

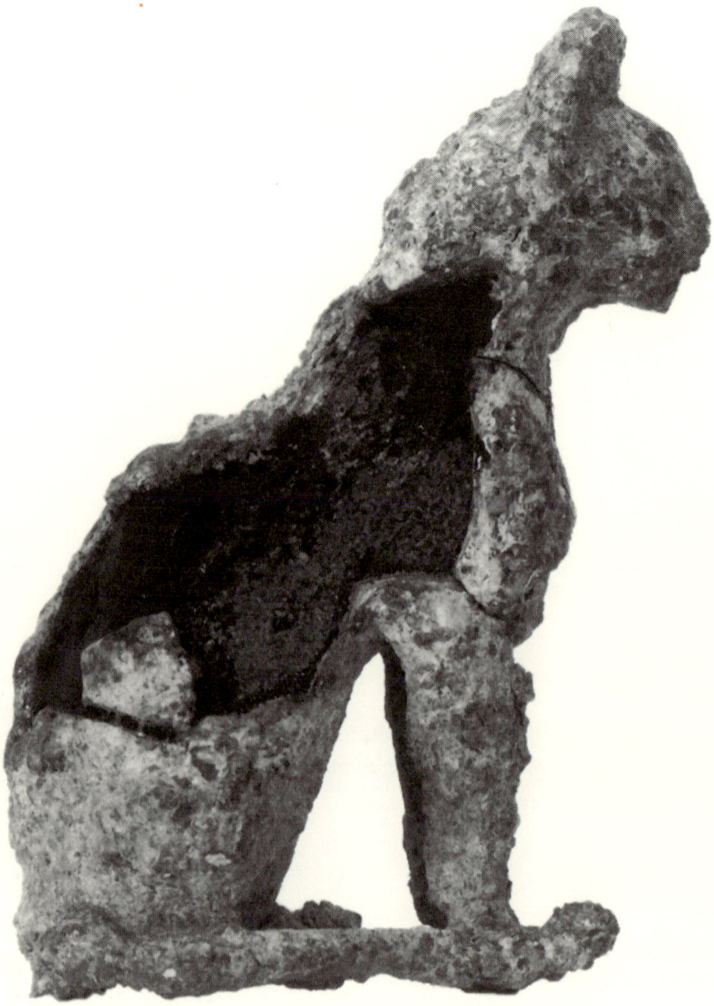

48 *Bronze casting of a cat, broken open and revealing a permanent asphalt core left in. Origin: Seventh century* B.C.*, Sakkarah, Egypt.*

into Greece, Rome and, finally, all Europe. Seeking copper and tin ores, and other metals, the metallurgists of the ancient river-valley cities traveled widely, and soon the metals of Spain, Britain, southern Russia, Brittany and Slovakia were fusing in foundry furnaces, to be shaped into weapons for warfare and tools for the production of wealth.

The art of casting copper appears to have been well established in the western Mediterranean lands by around 2500 B.C., and to have penetrated deeply into central and western Europe. By means of wax molding, the goldsmiths of Egypt produced jewelry that was the envy of the barbarians on the northern coast of the Mediterranean, and, due largely to their casting ability, Egypt was able to rule the western civilized world for nearly two thousand years.

49 *Cast bronze cat of type similar to casting of Fig. 48, but with core removed. Marks of core prints remain, indicating a considerable knowledge of foundry practice in early Egypt.*

At this point in history, even earlier than 2000 B.C., the Egyptians, or the people in northern Asia Minor, contributed still another and tremendously important chapter to the story of metal. They became acquainted with iron.

The importance and significance of iron was not at first generally recognized, mainly because the people had been so long primarily occupied with bronze, which they considered the metal of the gods. Lacking equipment to accomplish greater melting temperatures, they could produce only wrought or forged iron. Without the facilities for its reduction to liquid form, iron, like all metals, did not achieve greatest importance until it could be cast.

The reference books and histories bristle with controversy as to the exact date and manner of the appearance of iron. Without entering into this argument, it nevertheless is necessary to set down certain facts as established by the most reliable of archeologists and historians, since some of these data affect the history of castings.

It seems unquestionable that iron did not come from meteors. While attractive in theory, the hypothesis is wholly unsupportable. Meteoric iron contains a large percentage (usually above 5.0) of nickel, yet few pieces of forged iron have been found with anything but traces of nickel. Furthermore, spectacular as meteors are, their occurrence is relatively rare and, in view of the vast amount of trial and error required to develop any metal, it would hardly seem possible that the supply of meteors would have been sufficient for any extensive development of iron, although it is believed that such iron actually was worked.

Finally, meteors are extremely hard, due to their metallurgical structure. In 2000 B.C. and for a good many centuries thereafter, men had neither the tools nor the melting equipment for any appreciable working of meteoric iron.

The most reasonable theory as to how iron was discovered seems to be the simplest: that man deliberately attempted to melt iron ore and was successful. Recalling the ingenuity necessary to bring about the lost-wax molding process, and man's familiarity with melting by this time, it is not difficult to understand how this metal could have been charged into some early melter's crucible or fire and reduced to a bloom of wrought iron.

Another point concerns iron's relative newness in comparison with copper, gold and other metals of the non-ferrous family. Scattered traces of iron have been unearthed in Egypt dating back beyond 2000 B.C., but conclusive evidence cannot be found that, prior to 1500 B.C., such examples of iron were other than accidents. Iron was rare in the Near East up until the time of Christ. It was offered as prizes in the games at Sparta and was used for money and jewelry in both Greece and Egypt, thus proving its intrinsic value and indicating its probable scarcity.

It should also be remembered that the iron first discovered was not cast. The Greeks, about the sixth century B.C., learned from the Chinese that iron could be poured, and they cast some statues, weapons and implements. Nevertheless, iron was not cast generally until the fourteenth century A.D.

Development of the forging of iron was an important prelude to iron casting, just as the forging of copper was necessary to the complete use of that metal. Both copper and iron undoubtedly were smelted by the manually operated bellows as developed in Mesopotamia. When it was realized that the melting of iron required higher temperatures, the iron smelter built his furnace on a windy hillside so as to allow natural draft through increased stack height and thus assist the crude bellows arrangement.

The early hearths, somewhat similar to the later Catalan forges, used a brown hematite ore which contained less than 40 percent iron, with about 2 percent manganese. Blast was applied for about eight hours, during which time slag was drawn off occasionally, reducing the iron into a spongy lump or bloom, which then was split up into bars. This low-grade wrought iron then was forged into shape on reheating.

Iron deposits were found throughout the eastern Mediterranean area, principally in northeastern Asia Minor, where the deposits were worked by the Hittites about 1300 B.C. Iron is mentioned frequently in the Bible, which also contains many references to copper, bronze and gold. The workers in metal generally were classed as smiths, whether they were engaged in casting copper, gold or silver, or in forging iron. The impression has grown that all mention of early smiths concerned forging, but this is not the case.

In the twelfth century B.C. a vigorous race of people made their entry into the Mediterranean. These were the Phoenicians of the eastern coast line, who dominated the civilized world for five hundred years through their ability in trading and commerce. They were the first true merchants, and they sailed not only the Mediterranean but the Atlantic as far as Britain, as well, in search of tin, gold and other metals. Their trade in the products of the nonferrous founders and ironsmiths reached far into western Europe, eastward to India and southward along the Nile. Thus is illustrated the driving force of the metallurgical trend as a civilization builder.

The Phoenicians founded their chief commercial city, Tyre, in what is now Palestine, and then built Carthage in North Africa across from what was to become the most powerful of empires, Rome. Not only did the Phoenicians search for metals and spread their metallurgical and foundry knowledge, but they also produced the first foundryman whose name has come down to us—Hiram of Tyre, who is said to have lived about 1066 B.C. He is in part a legendary figure, yet some of his castings are known and described today.

Hiram's specialty was the casting of large bronzes of intricate shape and complexity. Perhaps his best known work was King Solomon's pool, 15 feet in diameter with a depth of 7 1/2 feet and weighing 45 tons. Originally the casting was mounted on eight cast bronze oxen; however, some 300 years later, the then ruling monarch sold the oxen, and the pool was dismounted. No trace remains today (Fig. 50).

This same Hiram of Tyre cast the pillars for Solomon's magnificent temple in the River Jordan valley. The pillars were 27 feet high and 4 feet 6 inches in diameter, with a total weight of 175 tons. It is interesting to recall that Hiram was greatly pleased with the Jordan valley because the clay found there produced fine brick and excellent crucibles for foundry work. Legendary though Hiram of Tyre may be, he was the first founder of record, and as such his works deserve greater recognition.

In the meanwhile, European civilization was being born in Crete and in Greece. Crete, which entered the age of metals about 3000 B.C. and developed a remarkable bronze casting industry, and the ancient copper-rich island of Cyprus, are the earliest lands mentioned metallurgically. It is fabled that in 1500 B.C. a forest fire accidentally showed the residents of Crete how to reduce iron ore to a liquid state. However, the facts seem to indicate that their knowledge of metal and how to work it emanated from the Phoenicians, from Egypt and from the Near East.

Cyprus is important because it is from there that copper derives its name. Copper was certainly the Cretan's most cherished possession, and they called it *Cyprium metal.* Under the Romans the name became *coprum* and later was Anglicized to *copper.*

Greece presently commenced her glorious development, and immediate enthusiasm for copper and its alloys was shown by

50 *Drawing of King Solomon's famous bronze pool, reputedly cast about 1066 B.C., the work of Hiram of Tyre. The bowl—15 feet in diameter, 7 1/2 feet deep, weighing 45 tons—was originally mounted on eight bronze oxen, cast separately from the bowl. No trace of this fabulous casting remains today.*

Grecian metallurgists. Already the alchemists had begun to experiment with all substances in their determined search for a method of producing gold synthetically. It is believed that the alloy brass came into being as a result of such experiments—that the alchemists made additions of zinc instead of tin to copper and thus obtained brass rather than bronze. The metal used was actually calamine zinc carbonate, although zinc was not recognized as a separate metal until the sixteenth century A.D. Melting of all these metals was carried on in crucibles over a bellows-blown fire.

The Greeks should also be credited with some metallurgical advances in ironworking. About the time of Homer (850 B.C.), the Greeks learned to temper iron by heating and quenching; at first the quenching medium was water, and later, oil. While the Greeks evidently knew how to cast iron, there is slight evidence that they practiced iron founding extensively. The name of one Greek founder has come down to us, along with some of his work—Theodomus the Samian, who probably learned from China the art of iron casting. He cast an iron statue of Epaminondas, an iron throne of Peridon and an iron statue of Hercules.

We also have the word of the great Aristotle, who lived to 322 B.C. The following direct quotation from the works of the world's first scientist offers additional proof that melting of iron was not general in Greece:

"Wrought iron itself may be cast so as to be made liquid and to harden again, and thus it is they are wont to make steel. But this they do not often because of the great waste and because it loses much weight in refining."

Aristotle indicated also that he was familiar with Indian crucible steel and the possibility of using coal as a fuel, although there is no clear evidence that either of these processes was used consistently. This again bears out the thought that the source of Grecian knowledge of metalworking was the Orient, where these methods were already in use.

As a monument to the art of the Greek foundrymen, the famed Colossus of Rhodes is considered one of the seven wonders of the ancient world. This enormous collection of castings was erected about 290 B.C. in the Dodecanese, on the island of Rhodes. It was a bronze statue of Apollo over 105 feet high, and weighed 360 tons complete. Cast in sections and equipped with a winding staircase, it stood with one foot on each bank of the harbor entrance. Although it was overturned and shattered by an earthquake in 224 B.C., this colossal casting has been classed with such man-made wonders as the pyramids, the Sphinx and the hanging gardens of Nebuchadnezzar (Fig. 51).

Greece, even with this wonder of the world, its fame for having established the first democracy and its highly advanced civilization, nevertheless was doomed by a new and powerful force in the Mediterranean (Fig. 52). Roman legions began to march the face of the earth and, one after another, nations fell before the force of arms (Figs. 53, 54, 55). First was Carthage in 146 B.C., burned and leveled by the Romans, in a war waged largely to gain possession of the metals which Carthage controlled in North Africa and in Spain and other European areas.

51 *One of the seven wonders of the ancient world—the Colossus of Rhodes—was an enormous bronze statue of the Greek god Apollo. Erected about 290 B.C., it towered 105 feet above the harbor of Rhodes, one foot on each shore. The colossus was cast in sections and assembled, weighed some 360 tons complete, and was equipped with a winding staircase. Shattered by an earthquake in 224 B.C., no trace remains.*

52 *Ceramic vase depicting the casting of statuary in Greece, about fifth century B.C. Decoration shows the master drawing metal from the furnace and (right) the assembly of castings into completed statues.*

53 *Roman limestone half-mold for casting bronze coins. Designs represent Victory, Fortune and Athena—all cast two-on.*

54 *Forged iron fork from early Roman times, unearthed at Mt. Vesuvius.*

55 *Forged iron sickle from early Rome, dating to about 100 B.C.*

56 *Cast bronze Roman pitcher, 77 A.D., found on the site of Bascoreale, buried by the eruption of Vesuvius that also destroyed Pompeii in 79 A.D.*

However powerful Rome became, it contributed little to the advancement of metallurgical technology and the foundry art. "Corinthian bronze" is typical of Roman progress. This material, an alloy of copper, gold and silver, is said to have been accidentally blended when the Romans burned Corinth.

War unquestionably aids in certain developments, but the Romans were primarily warriors and little interested in technical progress, except where it fitted specifically into their military plans. No attempt was made to advance, in their peacetime pursuits, the metallurgy demanded for their military purposes. The lack of progress possibly stems from the fact that slave labor was employed to a great extent.

This is not to suggest that Rome used no castings; on the contrary, castings were employed widely throughout Roman times (Figs. 56, 57, 58). One Roman circus contained more than three

57 *Child's face cast in the handle of pitcher illustrated in Fig. 56, showing excellent detail in spite of corrosion.*

58 *Bronze table unearthed from the area near Vesuvius. Note legs cast in shape of lion paws. Centerbrace casting appears to have been brazed to legs, while cast top is held in place only by its own weight.*

59 *Earliest known cast bronze bathtub from Roman foundries, cast in two pieces and brazed together. Ornamental lion-head handles were cast separately and apparently attached later.*

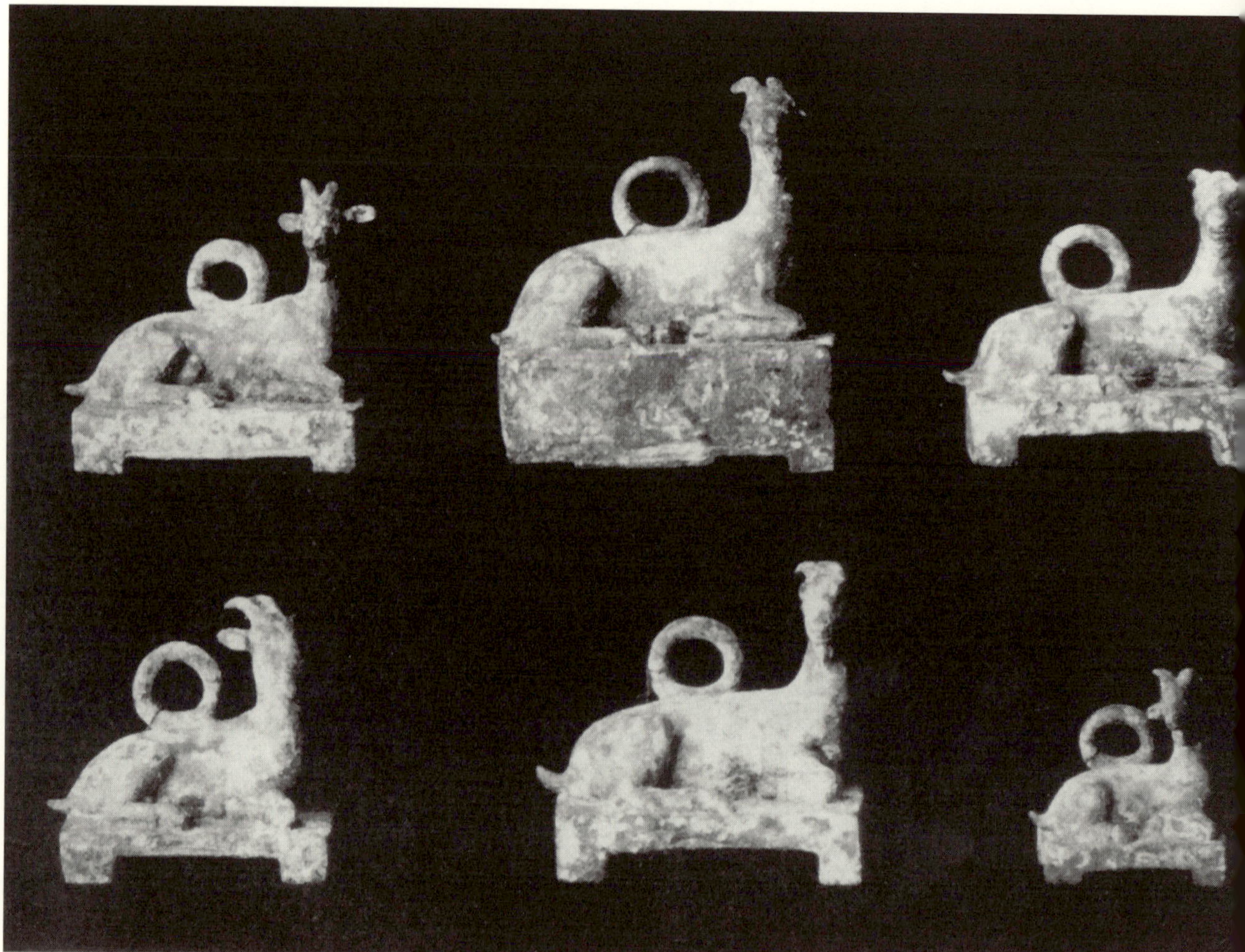

60 *Bronze scale weights (Roman), all cast in shape of goat. Largest weighs about ten pounds, smallest one pound.*

thousand cast-bronze statues, and every luxury in bathtubs, ornaments and household articles was indulged. However, Rome, unlike other early civilizations, contributed little or no original foundry progress; and as the empire deteriorated, so actually did technology.

The only ray of light in the gloom that immediately prefaced the Dark Ages was Pliny the Elder, Roman author of the first complete thesis on natural history. Pliny gave attention to metallurgy and described the iron, steel and bronze of those times (Figs. 59, 60, 61). His inquiring nature cost him his life when he approached erupting Vesuvius too closely during the destruction of Pompeii. However, Pliny was chiefly a reporter and added no new developments to the casting art.

61 *Other Roman weights, cast in various shapes: Left and right, Knucklebones; hog, at top, weighs 100 pounds; center casting in shape of cheese, with handle of two joined thumbs.*

Living by the sword, the Romans eventually perished by the sword. Nearly four centuries after the birth of Christ, the decayed and corrupt Roman empire, which had destroyed initiative, choked enterprise and bred despair and fatalism in its efforts to perpetuate power, fell before the onslaughts of the barbarian Huns from the north. And with the fall of Rome ends what may be called the third period in the development of the foundry, with works of great promise already accomplished in spite of pitifully inadequate equipment and a technology still in its infancy.

Darkness now set in, and the civilized world, at least from the point of view of science, entered upon a period of intellectual lethargy that lasted nearly a thousand years. The Middle Ages, or "Dark Ages" as we have come to know them, yielded few known advancements either in metallurgical progress or in human endeavors.

CHAPTER 4

Melting and Molding Methods of the Renaissance

DURING THE DARK AGES, following the fall of Rome in the fourth century A.D., whatever developments were made in European founding were almost entirely shrouded in mystery. The feudal system was in full swing in the western world, and competitive development, a necessary component of initiative and progress, was virtually strangled.

It should not be inferred that the foundry industry of the West did not survive the near-primitive state that Europe reverted to after the Roman legions withdrew. Founding did not actually become a lost art during the Middle Ages. Gaul, famous for its metalwork in Roman times, preserved considerable of the arts during the Merovingian dynasty. Charlemagne was a patron of metalworkers as well as of men of letters and lavished cast-bronze work on the church he built at Aix-la-Chapelle. In Ireland, foundrymen ranked with the nobility from early times, and the earliest Christian schools taught metalwork. The Anglo-Saxons also held their foundrymen-smiths in high regard.

Even so, the foundry industry of the West might have withered and died in the thin soil of feudalism and under the heel of the barbarian, had no other factor been present. The church and the rich and powerful feudal barons were the dominant influences, and technical secrets in the metallurgical field for the most part were carefully guarded as the sole property of the monks and clerics. The monastic orders became the refuge of many fine craftsmen, however, and as monasteries multiplied and their manorial holding increased, the number of foundrymen attached to the church was probably great. Likewise, the lords and knights had their own foundrymen-smiths in every manor and in every village.

Metallurgically, the chief activity was that of the alchemists, who were interested solely in the synthetic production of gold. Their theory was that gold was basic in all substances; for example, it was held that copper-tin alloy was an inferior gold rather than an improved copper. Alchemy sprang from the Grecian metalworker's conception of the coloring of metals, which directed attention mainly to surface appearance rather than to metallurgical structure. With such basic errors to build on, it is obvious why little of importance could be accomplished.

True, the building of churches, great abbeys and fortified works developed a need for metal and metal craftsmanship. For example, the demand for church bells resulted in marked expansion of the medieval castings industry, and aided the adoption of new molding practices and tools. The church also provided a growing outlet for many things in metal: fonts, croziers, candelabra, doors, censers, lecterns, to name only a few.

The chief bar to our determining what technical developments, if any, took place during those times has been lack of written data. Whatever information was available seems to have been passed almost solely by word of mouth, and practically no written word remains. The church again played the main role in circulating information. It alone had broad contacts and the latitude for travel; so it was naturally the main medium (and controlling factor) for transmission of new ideas and thought—and the main interest of the clerics lay in theology, not in metallurgy.

The one exception to this was the monk Theophilus, whose writings include descriptions of foundry practice in the eleventh century, apparently as a manual of instruction. His commentaries on bell molding indicate that he was one of those early practical foundrymen who "black'd his fingers and sing'd his beard in metallick operations." His *Schedula Diversarum Artium* is the earliest known foundry text. After Theophilus there is a continuous void in our records until the fifteenth or sixteenth century, when renowned foundry texts were prepared by the great Biringuccio, da Vinci and Agricola.

We are concerned less with writings and more with the results obtained as the European world began to shake off its lethargy and awaken from ten centuries of feudalism. Among these results were the advent of the blast furnace for iron, the use of the reverberatory furnace and other melting methods for non-ferrous metals, and the increasing emphasis on loam molding.

62 *Ladle used by the Romans during their conquest of England and typical of foundry ladles in use before Christianity began.*

In tracing the development of the blast furnace, the iron of those days should be clearly defined as being almost exclusively forged iron. Castings probably were produced occasionally, but even as late as the fifteenth century, cast iron was termed "corrupt metal," since it was believed that the liquefaction of iron ruined its properties. Considering the equipment and methods of the time, this quite possibly was the case.

No understanding of the chemistry of iron was apparent, and even the difference between iron and steel was not fully understood. The only known process of steelmaking involved reheating an iron bar in a charcoal fire until steel was obtained. This cementation process, of course, had been in use for centuries before the birth of Christ, but the liquid crucible steel of early India still was a lost art.

The famous Toledo steel was produced in Spain about this time by an unusual process. After steel had been obtained by cementation, the steel rods were rusted in the ground for a few years, presumably to remove impurities which the sword maker thought he recognized in the rust. Spanish steel, however, owes its sterling properties not to rusting but to the high-quality iron ore found in the Bay of Biscay region. With steel from this ore—exported to the United States even today—Rome had forged a mighty empire.

The making of iron and cementation steel was introduced into the British Isles about 1000 B.C., first by the Phoenicians and finally by the invading Romans, who remained there until 400 A.D. Under the Romans the industry seems to have been rather extensive, and after Caesar's legions departed, iron was smelted in Britain by Anglo-Saxon monks. A small cast statuette has been found in Sussex dated 170 A.D., but, as was the case on the Continent, cast iron did not come into general use until much later (Fig. 62). A monk, "the venerable Bede" (673-735 A.D.), mentions iron, and old records indicate the interest of the abbeys in producing the metal. The first so-called iron foundry in England is said to have been established by the monks of Krokstede Abbey in Lincolnshire, in 1161 A.D., but this undoubtedly was a forging operation.

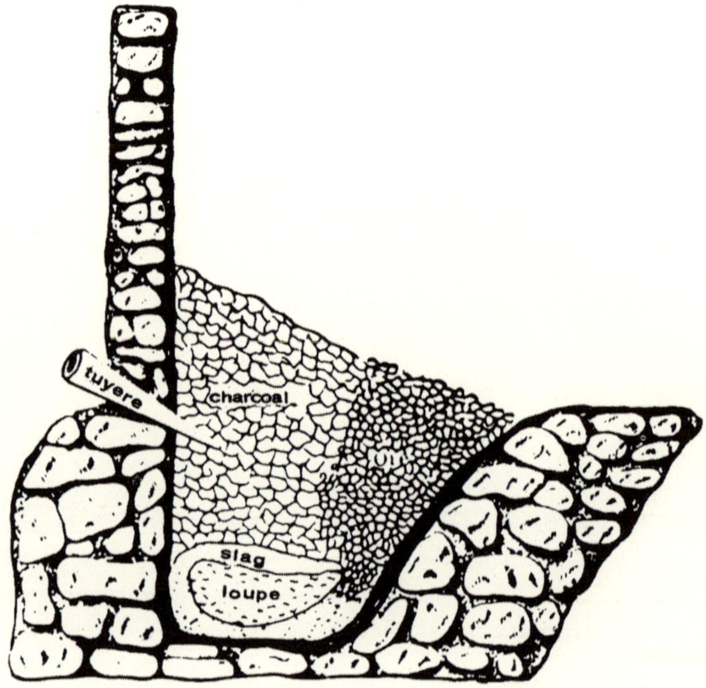

63 *The Catalan forge, developed by the Moors while occupying Spain during the eighth century A.D., produced wrought iron only. Note single tuyere. Ore and charcoal were charged vertically, resulting in a "loupe" or ball of iron, which was hooked out and hammered into a bloom.*

The true line of development of the blast furnace, which leads to the cupola, was not established until the eighth century A.D. by the Moors in Spain, through their now famous Catalan forges (Fig. 63). The Catalan forge produced wrought iron only. It consisted of a low hearth with one tuyere inclined at a 30-degree angle, ending 9 to 15 inches from the bottom. The hearth was about 40 inches square and about 16 to 27 inches deep. The forge took a charge of ore varying from 300 to 900 pounds. After a few hours on heat a ball of iron was formed, resulting in what was known as a "loupe," which was hooked out and hammered into a bloom. In turn, the bloom was split and shaped into bars.

In firing, the weight of the iron ore charged was approximately three and one-half times the weight of the charcoal. The relatively low temperatures attainable produced a spongy mass of iron, and the charcoal ash fused with the clay to form slag (Fig. 64).

Following this simple type of furnace came a Swiss model which differed from the Catalan in that the Swiss was built entirely above ground, the charges of ore and charcoal being made in successive layers, whereas in the Catalan forge the charcoal was placed in columns vertical with the ore. This Swiss furnace usually was built deep in the forest near charcoal reserves. The hearth was from 6 to 8 inches thick, with no foundation, and was constructed of firebrick. Walls were built of clay and stone.

A channel extending the entire width of the hearth was left open about two inches above the hearth. The shaft itself was about eight feet high, cylindrical, and inclined at the top so that the charge would pass down one side, leaving the door free from accumulations. No tuyeres were employed, since air entered the door by natural draft. The furnace, roughly illustrated in Fig. 65, usually produced blooms weighing from 30 to 50 pounds.

64 *How charcoal was made in early times. Piles of wood were covered with earth, lighted, and allowed to char for several days. (This and similar woodcuts are from an early edition of Biringuccio's,* ***Pirotechnia.****)*

65 *Old woodcut depicting the use of a manually operated bellows in smelting refractory iron ores. Installation typical of Swiss-type furnace developed some time after the Catalan forge.*

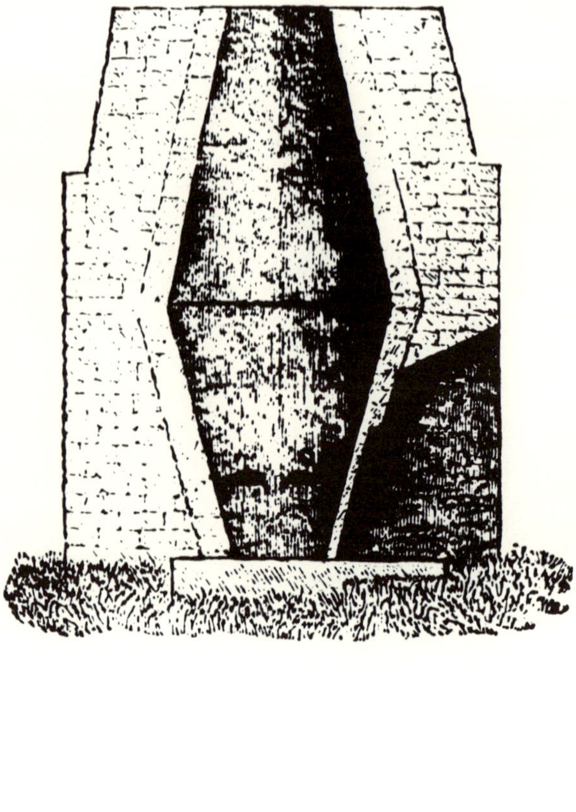

66 *The German Stückhofen, or wolf furnace, used in Belgium about 1000* A.D. *This furnace started the construction of bloomeries of greater height and stronger draft, leading eventually to development of the true blast furnace.*

67 *Osmund furnace, used principally in Sweden in the eleventh century, featuring increased height and one or more bellows-operated tuyeres. The usual ore was bog ore, calcined and charged. Note bloom being cut (top, at right).*

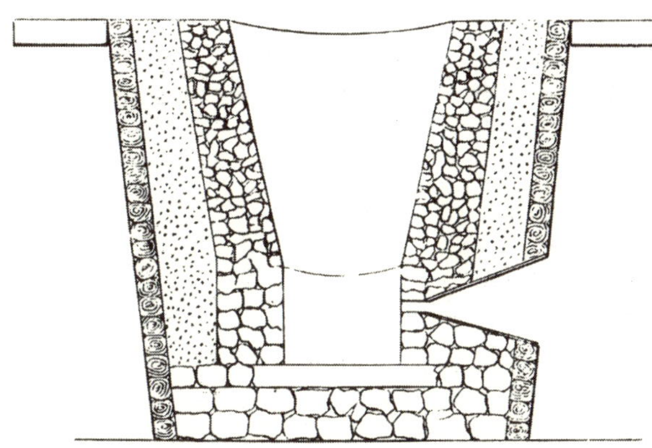

68 *Illustration of furnace operation, depicting the turning of a "loupe" of reduced iron.*

About a thousand years after Christ, two new types of iron smelters appeared and formed the intermediate link between forge and blast furnace. First came the Stückhofen, or wolf furnace, found in the valley of the Meuse River in Belgium (Fig. 66); the other was the Osmund furnace (Fig. 67), used principally in Sweden (Figs. 68, 69).

Main developments in Sweden centered around greater height (for increased draft and higher temperatures), and the use of one or more tuyeres functioning by means of manually operated leather bellows.

69 *Woodcut showing a smith of the Middle Ages, to whom most of the bloomery wrought iron was sent for forging.*

70 *Refractory ore furnace, with draft obtained by bellows arranged for direct operation by means of a man's weight.*

71 *Operating the blast bellows by means of a racking bar.*

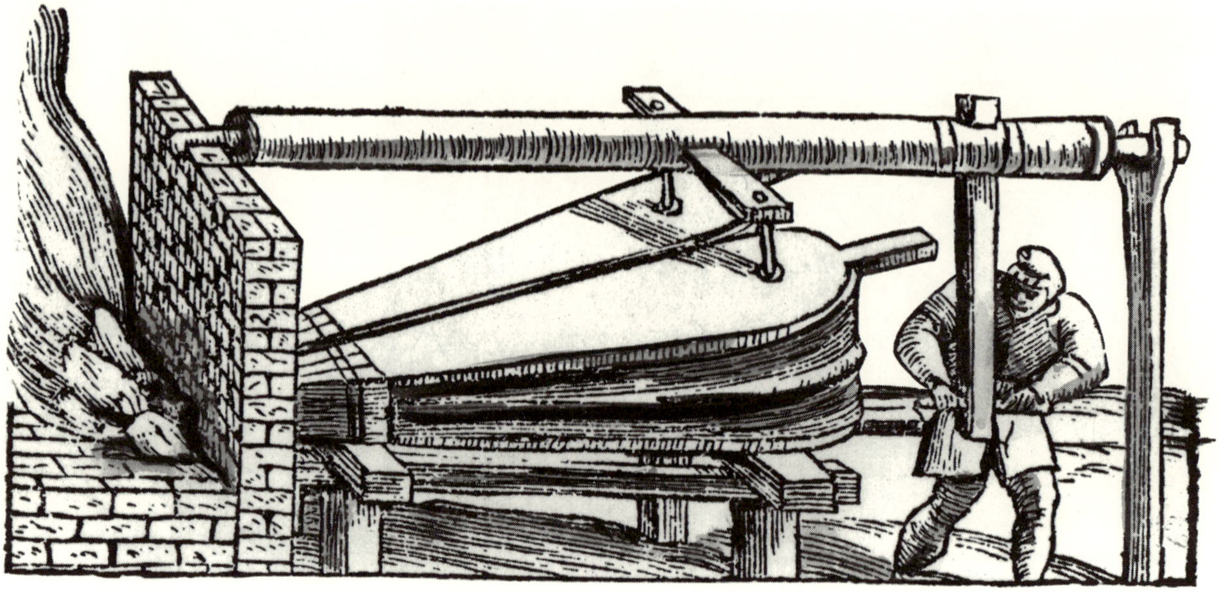

72 *Operation of a bellows by means of cams mounted on a shaft driven by an overshot water wheel. Molten metal, however, did not become available until advent of the water-wheel-driven bellows in 1325, producing the necessary draft for higher temperatures.*

These bloomeries of greater height and stronger draft were by this time approaching the molten stage, requiring only the addition of the water-driven bellows in 1325 and a slightly higher shell for the production of molten metal direct from the blast furnace—marking the actual beginning of a modern iron-foundry practice (Figs. 70-74).

Since charcoal alone was used as a fuel, carburization occurred, resulting in a brittle iron. The reason for this structural change was not clear at that time, but we now know that better iron was obtained during the reheat because less charcoal (i.e., less carbon) was required as fuel. In consequence, the principle of reheating developed slowly.

As time went on, blast furnaces were built larger and larger so as to achieve economical operation. Greater capacity permitted the metal to stay in contact with the fuel for a longer period of time, until finally molten metal was obtained.

73 *Front view of blast furnace operated (early fourteenth century) by a water-wheel-driven blast, showing (A) furnaces, and (B) forehearths.*

74 *Same furnace illustrated in Fig. 73, showing bellows and cammed shaft, driven by water wheel (not shown).*

75 *Types of early crucibles (l to r) scorifer, crucible, cupel. The word* ***crucible*** *derives from the Christian cross (crux) with which ancient alchemists adorned their crucibles.*

In the first quarter of the fourteenth century, temperatures were obtained high enough for slag to be removed and an iron rich in carbide, but hard and weak, to be run out of the furnace into shallow open troughs. The metal was tapped into a large reservoir and then into smaller and smaller molds, which to some imaginative smelter resembled a sow with sucking pigs. Thus was born the term "pig iron."

For nearly five centuries the iron foundry was to exist on direct-furnace metal without remelt, until the advent of the cupola and other developments opened up a great field of iron and steel chemistry, the full possibilities of which could only be realized when iron became castable (Fig. 75).

Other metal methods of interest, principally in the non-ferrous field, emerged in the fourteenth and fifteenth centuries. Outstanding among these advancements were refinements in the use of the crucible and ladle and the development of the reverberatory furnace. The crucible, as has already been seen, is of great antiquity. It obtained its name from the cross or "crux" with which the alchemists adorned it as part of the superstitious ritual of their semi-religious investigations.

In the development of the crucible over a period of many centuries we again find the potter's art closely bordering on that of the foundryman. It was early discovered that crucible failures were due in part to lack of uniformity, but principally to impurities such as

76 *Treading out the clay in bare feet, and molding the crucible—two steps that long linked the art of the potter closely to that of the foundryman. As late as two centuries ago, the iron foundry molder frequently was referred to as "the potter."*

77 *Natural draft furnace for melting in crucibles placed over a simple grill, a method employed in the fourteenth century.*

78 *Following the simple, natural draft furnace of Fig. 77, the manually operated bellows was used to provide a crude form of blast. Still later, the crucible was placed directly in the fire itself and bellows-blown for draft.*

79 *Melting on the hearth and direct in the ladle by means of manually operated bellows.*

stones or other foreign materials in the clay. The centuries-old method of obtaining pure clay consisted of treading it with bare feet, which broke up the lumps and made it somewhat homogeneous. The treaders used a combination of stamping and twisting which brought up the strength and reduced impurities, and clay thus worked was used in the making of crucibles (Fig. 76).

Crucible melting was carried on in many different ways. In one of the simplest methods, ore was placed in a crucible set upon a simple grill and reduced over a natural draft fire (Fig. 77). Later on, a manually operated bellows was used to create a crude form of blast, as illustrated in Fig. 78. Still later, the crucible was placed directly in the fire embers and bellows blown.

From the small crucible developed the foundry ladle for use in pouring larger vessels. At first, metal was melted by blast and run into a hand ladle (Fig. 79); then the metal was melted in the ladle

80 *Melting in clay-lined basket-type ladle and on the hearth, both by bellows draft, and with bottom pour. The metal is contained by means of a clay plug, punched out for pouring. From such crude beginnings of the fourteenth and fifteenth centuries (Figs. 77-80) developed many pieces of equipment common to present-day foundry practice.*

proper, fuel being added direct. Finally, basket-type ladles and shallow open furnaces with bottom-pouring devices were brought in. Metal was contained by means of a clay plug, punched out when the operator was ready to pour (Fig. 80).

In all of these devices can be seen the crude beginnings of many pieces of equipment which today are used in the melting of all types of metal. However, these innovations must have created much speculation and comment among the foundrymen of the fourteenth and fifteenth centuries.

The earliest types of reverberatory furnaces were recorded first by Leonardo da Vinci and then by Biringuccio, both Italian engineers and foundrymen. No two furnaces were exactly alike, varying considerably according to the ideas of the foundryman who used them.

The principle of heat by reflection was generally known, however, and all types of furnaces bore some similar characteristics.

The early reverberatory furnace usually was a square furnace of firebrick, the flame from the fire-box or fire-boxes in the rear reflecting heat from the roof onto the metal in the hearth. Metal or ore and fuel were charged from the top, and the melt could be stirred and tested by means of bars inserted through openings in the sides. The metal was poured off at the front into a reservoir, whence it was ladled or channeled into the waiting molds.

This type of reverberatory furnace was the first outstanding development in metal melting since the days of the simple bellows-and-crucible units of the Egyptian and Greek founders and the box-bellows furnace of the Chinese (Fig. 81).

In an analysis of foundry conditions at the beginning of the fourteenth century, it is equally important to mention the improvements made in molding practice. Even before the birth of Christ, extensive use had been made of the lost-wax process, wherein the work of the wax modeler, the potter and the molder proceeded along parallel lines.

Such was the foundry art until the advent of loam molding, which gave the foundryman much greater flexibility. For a descrip-

81 *Plan of reverberatory furnace having two fireboxes. Furnaces of this type were first recorded by Da Vinci and Biringuccio in the sixteenth century.*

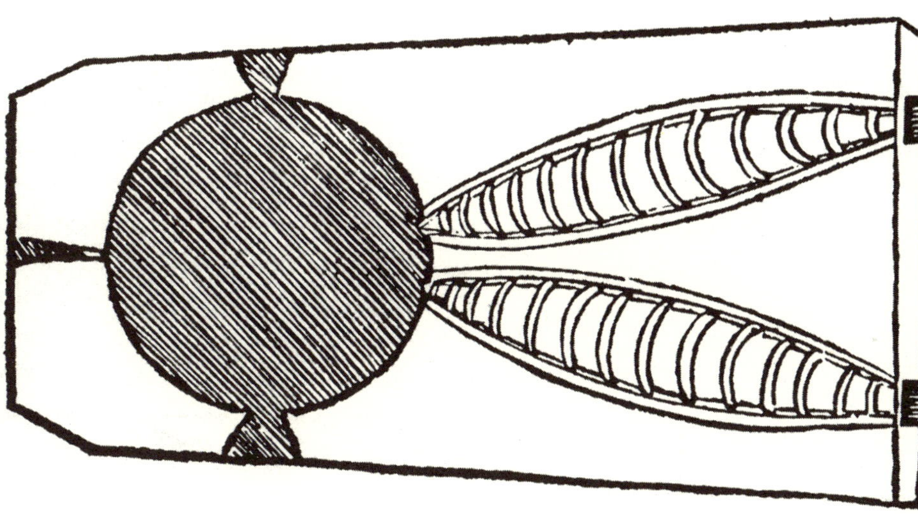

tion of this process, a great deal can be learned by direct quotations from that first true foundryman, Vannoccio Biringuccio (1480-1539). This man will stand for all time as the "father of the foundry industry." In 1538 he became head of the Papal foundry in Rome. The first man known to have set down foundry practice in great detail, his clarity of analysis, together with a common-sense, practical approach, mark him as an accomplished artisan.

To be convinced that Biringuccio knew the foundry business, it is only necessary to examine his statement that achievement in the art of casting "lies in three principal actions. One is making and arranging the moulds well; the next is smelting and liquefying the materials of the metals well, and the third is making the composition of their associations according to the results you wish to have." His *Pirotechnia* undoubtedly covers all that was known of metallurgy in the sixteenth century.

Unfortunately, his *Pirotechnia* has only recently come to light, and other and lesser developers have received the credit for much of his original work. Georgius Agricola, for example, whose name is familiar to every engineering student, states in his book on metallurgy that he refreshed his memory by referring to Biringuccio. (His "refreshing" consisted of copying, without credit, many of Biringuccio's most important chapters.) Perhaps some day proper credit and honor may be done this earliest of articulate foundrymen, some of whose passages on molding are of particular interest.

Describing the preparation of a loam suitable for casting "bronze, brass and other metals," Biringuccio goes into detail on clay sources and how to recognize molding clay. He then states:

"When you are to begin the work, you must choose the earth which you judge or which you have found by trial to be the best. If you cannot have it by itself, mix or compound it with another, as I have said. If it is too unctuous, put in some that is lean; if it is too lean, add some that is unctuous. Thus temper it to your own way until it becomes of a good quality.

"In order to compound it, put it on a bench in a heap and make a paste by moistening it. Then, just as potters do with theirs, proceed to beat it well with an iron rod and also mix it with two-thirds of the whole quantity of wool-cloth clippings. Thus, beating it again and again with the rod until you see that the clippings and the clay are so well compounded and united together that they appear to be one and the same thing, and the pebbles which perchance were in it are all very well crushed, and it has become soft and tractable, then proceed to make use of it in making the moulds, for this is the ordinary clay which you are to use. (Ref. Fig. 82.)

82 *Compounding loam by pounding with rods, as in Biringuccio's time, and preparing the mold in boxes, the forerunner of the modern foundry flask. It is interesting to note that the sixteenth century riddle (left) still is used extensively in many foundries today, some 400 years later.*

"There are some who, because they do not have or cannot find a clay that is as perfect as they would wish it, necessarily take that which they can have. First they make a paste of which they make cakes, and then they dry it, crush it when dry, then sift and again moisten it, fill it with cloth clippings and beat it.

"There are some others who mix various earths with it; others, wash ashes; and some, coarse sand. There are some who, having a weak clay, moisten it with salt water and add rust or finely ground iron scale, and some add crushed brick, especially in the first clays that they apply.

"Loam is accompanied not only with wool-cloth clippings but also with wool cardings and with those edges that fulleries take from cloth, as well as with hairs that are taken from skins in tanneries. But in fine or close works where the bronze must be thin these are not good because they are too long. It is also prepared with dry dung of horse, donkey, or mule; some with cow dung, some with chopped flax tinder, some with cane flowers, some with finely cut straw.

"And to conclude, in short, there are many things that would serve as an admixture for this clay, but the best for all that I know are the clippings of wool cloth. But because you sometimes find yourself in a place where there are not any, it is necessary to do what you can. Therefore, I have noted above all the things that can be used if there is nothing else. Sometimes the works force one not to make clays with cloth clippings; for example, the cores of many works that need a fragile clay so that they can be easily taken out of the inside of your casting, which cannot be done with those made of cloth clippings."

Another interesting reference from Biringuccio concerns general molding practice and methods of preparing molds for bronze castings. The following passages are quoted verbatim from his writings:

"The pattern should be a body of a harder material than the thing you wish to mould with and of a substance that is not swollen or disintegrated by moisture. Its archetype may be of marble, bronze, lead or any other metal, or also of wood, wax, tallow, gypsum, sulphur, or other plaster compositions—in short, of whatever you may have or whatever succeeds best for you, depending on whether the works are larger or small and easy or difficult to mould. If you wish to cast them in bronze, you must mould them with loam, covering them with it and then drying it.

"There are some things in which it is enough to form only one part, such as half- or bas-reliefs. These are easy if they have no undercuts. Some other patterns that are easily handled are formed in two halves, or in three or four pieces, if they are of a hard material.

"There are also moulds for large statues which, if one desires to make them of bronze, are first made of wax according to the ordinary procedure . . .

"There are likewise some moulds that are difficult to make, like scenes on a plane to which figures in full relief are attached, or twisted ornaments or other detached things that cannot be withdrawn from the mould without breaking either it or them. For this reason it is necessary to fill up again those places that hold fast and to make the mould of several pieces, with joints and good mortises made in such a way that, when taken apart and marked place for place, all may return again to their positions."

Another passage from Biringuccio has to do with molding in frames or wooden boxes with "powders":

"Small things are customarily moulded in two ways (unless they have undercut parts which hold them in the moulds): that is in clay, or with natural or artificial earthen pattern made in halves in either wooden boxes or bronze frames. This also depends on whether the object is larger or smaller.

"Now, if you wish to mould with moist clay, you must grease your relief with oil or pig fat; apply a dusting of charcoal, ashes, or cuttlebone on it; or gild or silver it dry with gold, silver, or tin foil. Then, first make with moist and somewhat firm clay a plate as thick and large as can conveniently hold your relief, and carefully set the latter halfway into this and let it dry. Then greasing and dusting it again, make the other half.

"When both are well dried, take out your relief and at the top make its gates and vents. If plastering is needed, do it. Finally, bake the halves, fit them together, and tie them well, as you wish, using the procedure that I taught you in the discussion of brass.

"But if one has to make a large quantity of works of one kind, he should for the sake of convenience use the method with powder because it is short and requires less time and expense. Therefore, if you wish to do this, you need to have several pairs of bronze frames or wooden boxes, higher than half the relief that you wish to mould, and of convenient length.

"On a flat board always keep half of one of these frames full of powder that has been slightly moistened for moulding, as I told you, so that by pressing with the ends of the fingers and with the hand - it holds together as well as possible. Level it off with a cutting tool and when leveled put a flat board over it and turn it over. Turn it on to a dusting of ashes so that it will not stick, dust it on the other side, and remove the excess by blowing it off.

"Then mould the thing that you wish to mould, pressing it halfway in once or twice, always taking it out and putting it back again. Then the other frame is fitted in its place and filled with earth by pressing and squeezing it carefully. Then it is leveled off on the outside with a sharp tool as you did with the other part, and then one of the parts of the frames is lifted up with the point of a knife or some other tool and, when opened, the thing that you moulded is taken out.

"If this is as well moulded as you wish, do nothing more; if it is not, replace it. In case some little piece should come off in lifting it out and you did not wish to have to remove it in the bronze, put it back carefully moistening it somewhat with salt water, white of egg, water of gum arabic, or some other viscous water.

"Then having made the gates or vents on this, if they are not arranged to be moulded at the same time as the objects, put the moulds upright to dry near a fire. If they are things that need a core inside in order to make them hollow and light in bronze or metal, such as pedestals, candlesticks, small or large bells, mortars, or similar things, make the core on a suitable iron, using either the same powder in a mould, or forming it by hand or with a strickle board from moist clay that is half or more ashes.

"Dry and bake them well and then put them in their places in the spaces in the frames as their marks will show you. Then smoke them well with the flame of a tallow candle or with that of a little turpentine. Then join the frames together and press them between two flat boards with a clamp or with a rope or in some other way so as to hold them firmly tight. Having brought them to this point, finally cast them in whatever metal you wish."

The new era of foundry development which opened with the end of the Dark Ages was well under way by the close of the fifteenth century. The blast furnace had been established and pig iron introduced, pointing the way toward iron and steel moulding. The various developments in non-ferrous melting—crucibles, ladles and furnaces—were ready to take their proper places in the foundry picture. Molding practice had advanced into loam molding, and the way was open for the use of sand and permanent patterns.

Progress, while slow, had been fairly steady. With greatly improved melting equipment, better molding practices, new efficient tools and the ability to cast both ferrous and non-ferrous materials, the foundry industry had fitted itself for the important industrial role that lay ahead. True, it was not yet a highly scientific industry; but it had the indispensable realization that the art of casting, as Biringuccio put it, "is dependent upon and subject to many operations which, if they are not all carried out with great care and diligence and well observed throughout, convert the whole thing into nothing."

CHAPTER 5

Casting of Bells and Guns

ALTHOUGH human progress, virtually stagnant for nearly a thousand years, did not resume its unfettered advancement until the beginning of the Renaissance period in the fourteenth century, it had been accumulating strength throughout the many centuries of the Dark Ages. Foundry practice was no exception. The manufacture of bells and guns—foundry products—typified the new forces at work, and new techniques of molding and casting these products placed the metal-casting industry in the lead as the world reached out toward new horizons.

Among the workers in metal who preserved foundry traditions and kept metallurgical knowledge alive in Europe were the bellfounders. During medieval times they were not specialists, but, rather, capable foundrymen-smiths who turned out weapons and armor, buckles, bits and bridles, household utensils and tools, swords and other implements demanded by the feudal lords. It was the church, however, that provided the greatest outlet for their skills in bellfounding.

Religion was one of the greatest influences during the Dark Ages after the fall of Rome. Construction of cathedrals and churches of tremendous size aided greatly in keeping the skill of the artisan alive. Integral parts of these edifices were the bells, the making of which was entirely dependent upon the founder's art. As cathedrals grew in size, the bellfounders supplying them were forced to cast larger and larger pieces, and because of the necessity of casting all in one piece, melting and molding methods were developed which proved valuable in casting later products (Fig. 83).

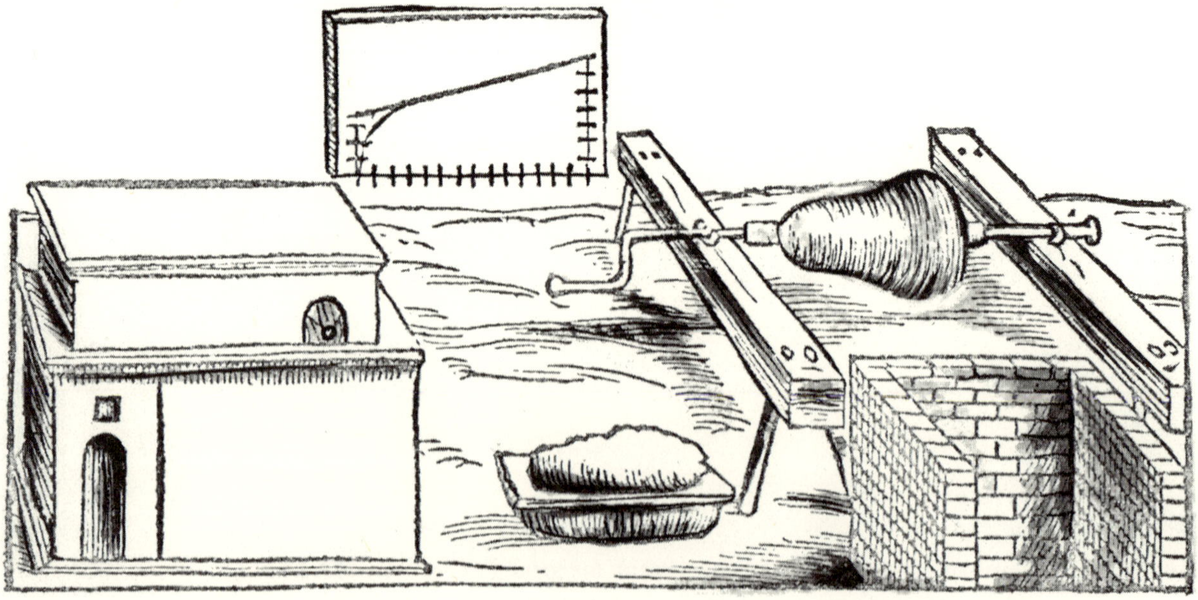

83 *Drawing of early bell foundry, showing (left) reverberatory furnace, (rear) bell scale similar to that invented by Biringuccio, and (right) horizontal sweeping of a bell core.*

The founding of bells undoubtedly was responsible for the development of loam molding and melting in the reverberatory furnace, two improvements which were to take the founding of metals out of the class of a purely artistic medium. Many of our important industries have undergone an artistic stage, but only the truly great industries have gone on to become useful in a fully utilitarian sense. The casting of bronze bells exerted just such an influence on the foundry. Because of its size and importance, bellfounding raised the casting of metal into the class of a practical art.

Bells were cast early in history. It is recorded that cast golden bells adorned the temple of Solomon, and hand bells of small sizes are referred to in the Bible. The first true tower-type bell was developed by Paulinus, Bishop of Nola, in Campania about 400 A.D.

Up through the eleventh century the bellfounders developed their knowledge, and from their practice emerged the first real tools of molding—the strickle board and sweep molding, the latter a process still familiar to present-day foundrymen. It cannot be said that the use of the strickle or forming board was a revolutionary step, but it was one of many far-reaching improvements in the technique of molding (Fig. 84).

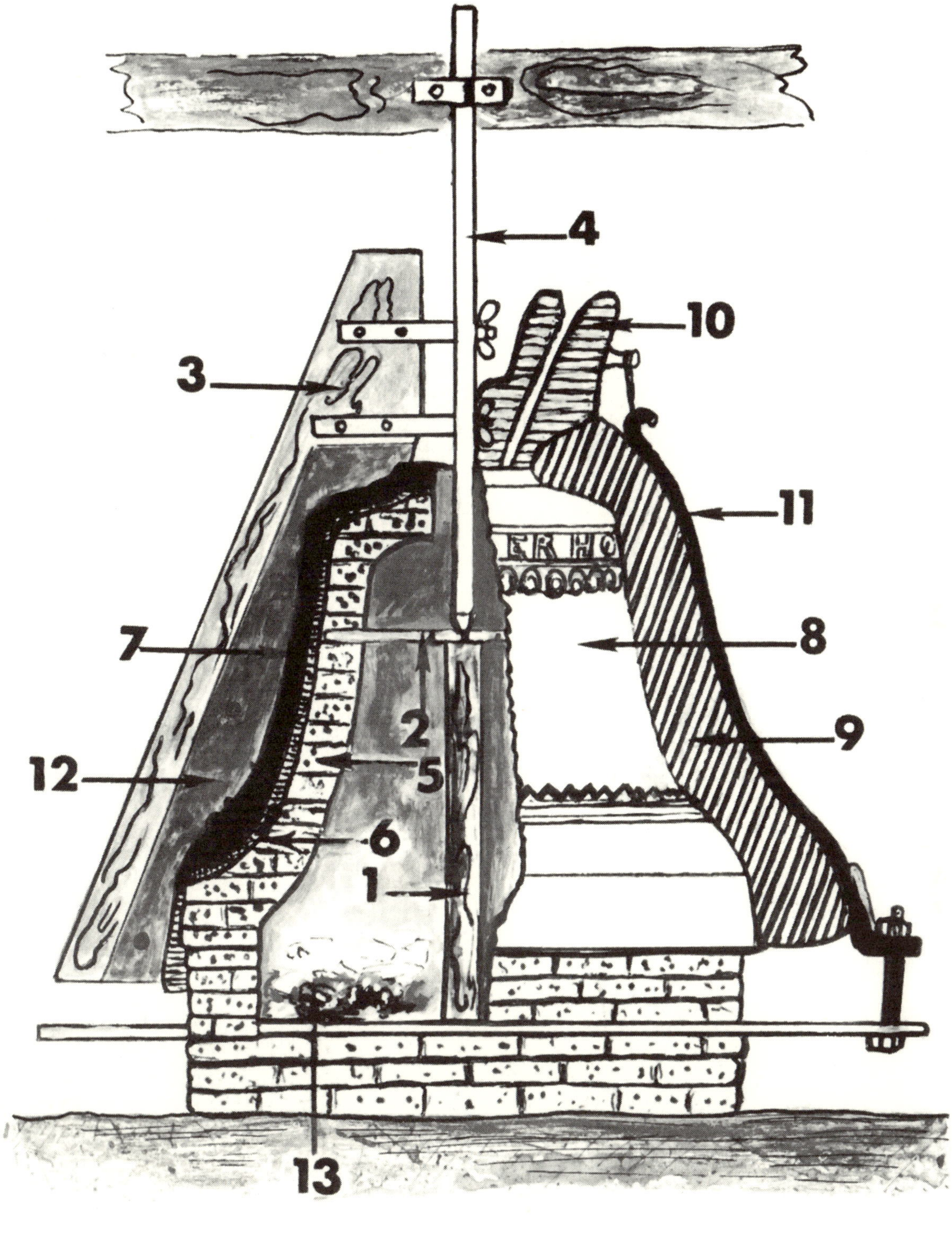

84 *Diagramatic scheme for sweeping a bell mold with a vertical strickle board. 1 arbor brace, 2 bearing board, 3 strickle board, 4 spindle, 5 brick work of core, 6 loam core face, 7 clay and wax bell pattern, 8 bell pattern, 9 cope, 10 pouring gate, 11 supporting rods, 12 metal profile block, 13 baking fire.*

85 *Sweeping cope of bell mold in pit. Note complete core and cope, at left. Vertical sweeping usually was employed in molding bells, coupled with pit molding.*

Bells today are molded to some extent in the same manner as in the early days. Many of the old practices have been continued, although vast improvements have been made in techniques and results.

Bellfounding also widened the scope of other foundry practices—for example, pit molding. Bell molds were formed in deep pits dug directly in front of the furnace, thus simplifying the pouring process. Molds too large for convenient floor molding are still made in pits today.

Bells further defined the distinction between cope, drag and core in the foundryman's mind (Fig. 85). The process called first for

86 *Finishing and touching up the cope of a bell mold preparatory to pouring the bell casting.*

87 *Assembling the cope and core of bell molds in an early bell foundry. Note reverberatory furnace in background.*

preparation of the core, formed either on a horizontal or a vertical spindle. Although records indicate that both types of sweeping were employed, it seems that vertical sweeping was usual in bell molding, while horizontal sweeping was the usual method employed when the bellfounders turned to the molding of cannon.

Vertical sweeping consisted of a top and bottom bearing, the latter supporting a spindle on which the strickle board was mounted. The board was shaped to the exact interior contour of the bell. Loam, based on a brick interior, was plastered on until the board could be revolved with clearance. This core then was coated with tallow as a parting and a new board was placed on the spindle, shaped to the outer contour of the bell. Clay again was added until the outer shape was attained, and this also was coated with tallow. Inscriptions were applied in the wax and more loam added to complete the mold.

Rods were placed in the mold for reinforcement and the entire mold baked. The mold then was completely disassembled, the clay pattern between the partings broken out, and the mold reassembled after finishing and touching up, as illustrated in Figs. 86, 87.

When the bells were cast, the metal was troughed direct from the furnace into gates, and after cooling the castings were broken out. There followed a breathless moment when the foundryman waited to see if his weeks of work were successful or not, for the entire success of bell casting was dependent upon tone.

True, a certain amount of chipping could improve the tone when the bell was tuned; but the bellfounders, unwilling to leave tone to chipping or chance, sought for the relationship between tone and the size, shape, metal composition and thickness of a bell. Thus the standard bell scale came into being—perhaps the first instance on record of the foundryman and the engineer combining their skills for the production of perfect castings. Biringuccio, again the diligent engineer-foundryman, was responsible for this scale of metal thicknesses, which still is employed in the production of cast bells (Fig. 88).

Thus bells, the great contribution of the church to the foundry art, were responsible for two important and progressive steps in foundry progress, aside from loam molding and reverberatory furnaces. The greater exactness called for in bellfounding brought about use of a precision tool (the strickle board) in molding, and established a relationship among design, metallurgy, foundry practice and the final product—a relationship we refer to today as "casting engineering."

Aside from the technical aspects of bellfounding, some of the great bells of history are worthy of attention. Italy, the birthplace of the art, sent the first bell to England in 608 A.D., where it was mounted in Wearmouth Abbey. The first bell cast in England was produced in 923 A.D. for Abingdon Abbey, and by 1150 bellfounding was established as a regular trade. Alwoldus was one of the earliest of English bell casters.

The trade flourished in England and throughout Europe, but in ways that differed somewhat from the conventional foundry business. The bellfounder's shop was usually the churchyard itself, where furnace, pit and mold were built. After the bell was cast and prior to its hanging, an elaborate ceremony usually took place

88 *Tuning a finished bell by means of a violin in an English bell foundry (ca. eleventh century). Development by Biringuccio of a standard bell scale is one of the earliest instances on record of the foundryman and the engineer combining their skills for the production of perfect castings.*

89 *The famous bell founders' window in York Minster Cathedral England, installed in 1320. Successive panels from left to right depict molding, hanging and sweeping of bells.*

wherein the inverted bell was filled with liquor for all the parishioners. We may well imagine that the founder and his helpers had by this time developed quite a thirst.

The importance of the bellfounder in England is depicted in the beautiful bellfounder's window in York Minster. The window (Fig. 89) was installed in 1320 and shows the three steps of molding, casting and hanging a church bell. It suggests, too, that the bellfounder stood high in the esteem of his community. England cast many famous bells, some of which are listed below so as to give some idea of the size of the foundry task they involved:

Location	*Name*	*Weight*
St. Paul's	Great Paul	17 ton
Westminster	Big Ben	14 ton
York Minster	Great Peter	13 ton
Beverley Minster	Great John	8 ton
Oxford Cathedral	Great Tom	8 ton
Exeter Cathedral	Great Peter	7 ton
Lincoln Cathedral	Great Tom	5 ton

One further sidelight on English bell casting illustrates the competitive spirit that flourished among founders even at an early date. As congregations grew, old churches and bells were pulled down and the bells recast by adding more metal. This situation occurred in 1742 at Stanford where a Rigby bell was recast by one Abel Rudhall, who lettered on the new bell the following legend:

"Badgworth Ringers they were mad
Because Rigbie made me bad
But Abel Rudhall you may see
Hath made me better than Rigbie."

While England and Italy probably gained greatest fame for their cast bells, thousands upon thousands of bells sounded throughout Europe and the Christian world before the close of the Middle Ages, good evidence of the activity and numbers of foundrymen. During the fourteenth century in England, bells were cast at a halfpenny a pound, and when the big struck bells were supplanted by swung bells, the founder frequently took his pay in the form of excess metal.

Russia produced the giant of all bells, the Monarch of Moscow, which, hung in the Kremlin, required some three years for molding alone. Four reverberatory furnaces were built to melt the metal. The completed casting weighed 220 ton and cost $360,000. Unfortunately, it fell and was broken in 1737.

Russia also is known for several other large bells; one at Trotskee weighed 171 ton, and another Moscow bell weighed 110 ton. When it is realized that these were single castings, capable of chiming, the marvel of the workmanship and foundry practice involved can be more fully appreciated.

Bells elsewhere in Europe were on the more conservative side. One cast at Erfurst in 1497 totaled only 14 ton, and another cast at Cologne fifty years earlier weighed 11 ton. One French bell, at

Bayeaux, was cast in 1202 and still is in daily use. Iron bells were produced in Bavaria in the latter half of the fifteenth century, and the trade was active in Berne, Berlin and Vienna in the seventeenth century.

The trademark of one fifteenth-century foundryman of England was a shield bearing a replica of a bell and a cannon. Many other bellfounders of Europe had turned gunfounders by that day, the first cannon having been cast in 1313 in Ghent. The bell-and-cannon trademark, however, represented an interesting and important expansion of the foundry industry and bore out a point made by Biringuccio—that the art of casting is a "necessary means to very many ends." For our purposes, bells and cannon strongly influenced the greatest expansion of the foundry industry in medieval times.

To the foundry industry, the advent of gunpowder brought completion of that process of expansion which began with the casting of the bell. It would be pointless here to develop the history of ordnance, since thousands of pages have already been written on that subject. However, the casting of guns exerted a profound influence on the progress of the foundry industry.

The first guns were cast in bronze in the year 1313, by a monk in the city of Ghent. These first guns were cored bronzes of two types: the bombard, a short mortar type piece; and the serpentine, of smaller caliber but longer length. It was from the latter type that the final iron artillery piece using cast shot was developed. At first guns were of the large bombard type, strapped to low benches and having low trajectory and little accuracy (Figs. 90, 91).

At this time a struggle in design had developed between the proponents of cast guns and those favoring a fabricated piece made up of wrought-iron bars bound together. Increased charges of powder and the demand for a greater number of guns finally demonstrated the superiority of the cast product, which prevailed for four hundred years with no marked change except for the general substitution of iron for bronze, about 1550.

The earliest projectiles used were darts, which gave way to concrete balls with a range of about one hundred yards. As early as 1373, however, iron shot came into use, albeit sparingly because of high cost. In the early 1500s the explosive shell or ball was first employed, and then the race among canister, grape, shrapnel and other death-dealing missiles began in earnest. Castings had little to do with most projectiles, with the exception of the conventional cannon ball, which will be described later (see p 114). As a matter of fact, stone cannon balls were in general use up to 1568.

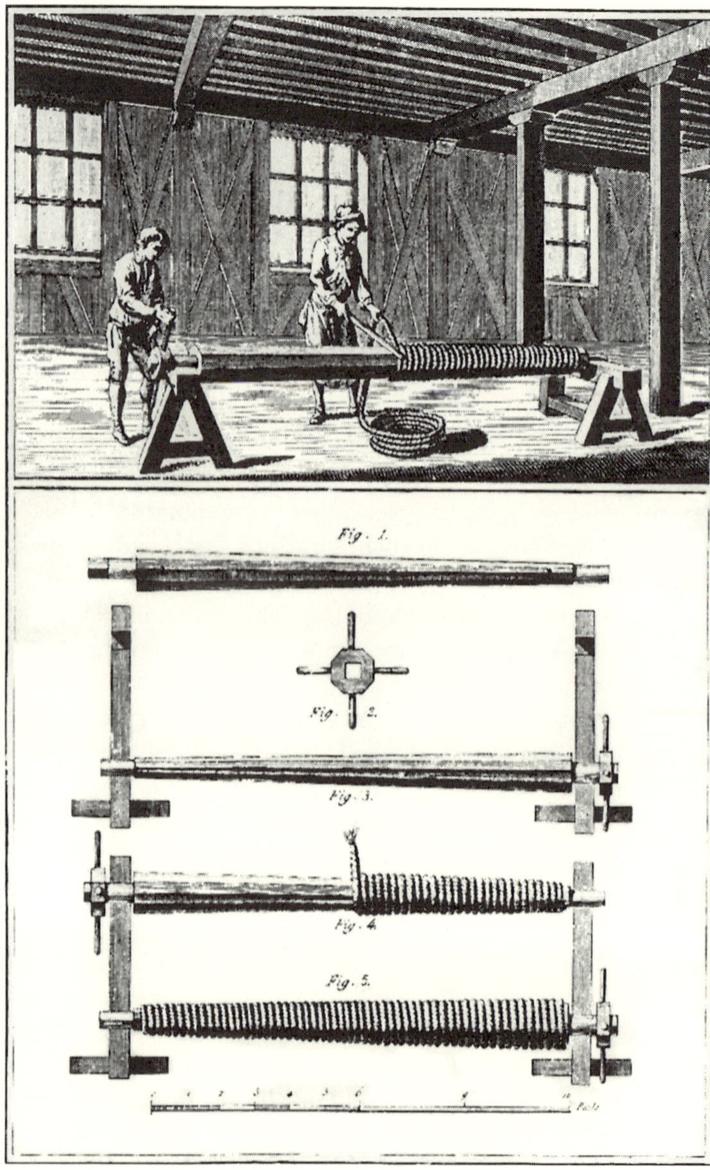

90 *Above: Wrapping the arbor with rope for the molding of guns (about fifteenth century). Below: (1) Arbor. (2) Winch. (3) Arbor and winch mounted on trestles. (4) Partially wrapped arbor. (5) Completely wrapped arbor.*

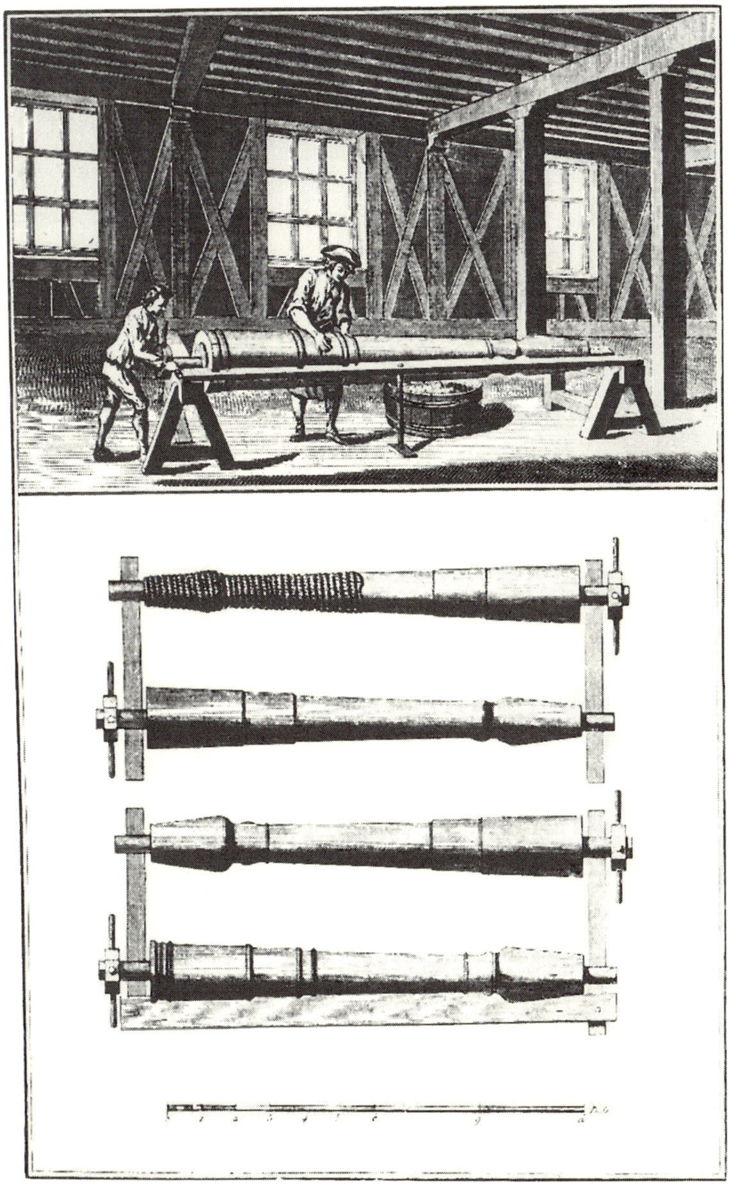

91 *Above: Applying loam and using the strickle board to obtain the pattern of the gun exterior. Below: Partially wrapped arbor, and successive stages in formation of gun exterior.*

Just as many early church bells were cast in the churchyard, many of the guns of medieval times were cast on the battlefield by the bellfounders turned gunfounders. The foundryman usually produced his cannon before a besieged city, a practice probably due to the difficulties of transporting heavy pieces of artillery. Such was the case in the siege of Constantinople by the Turks in 1453, the first large-scale use of cast guns for siege purposes. Such guns, however, had been used previously—on the Baltic Sea in 1350 and in the siege of Claudia Jessa by the Venetians in 1366.

The first famous iron gun was the Basilicas, cast in England about 1500. From that time forward the English iron gun was made in tremendous quantities. As a matter of fact, England herself made most of the guns, numbering over two thousand, with which the Spanish Armada attempted to conquer England. Henry VIII of England was greatly interested in cast-iron guns, and when the trade became unpopular at home, cannon were smuggled in. Early engravings place many of Henry's cast cannon at the siege of Boulogne in 1544. It is interesting to note that gunfounders received a nice price for this work, one record indicating that iron guns were sold for about twenty-five cents a pound.

Engaging as is the history of ordnance, this account is concerned mainly with the technical developments resulting from the casting of bronze and iron guns. In general, the molding of cannon was quite similar to bell molding, but certain improvements were carried forward from the bell era. Three are especially noteworthy: first, the use of the arbor upon which the whole mold was built; second, advances in core practice and, even more important, the extensive use of chaplets; third, greater realization of the fact that casting success depends largely on the gates and risers for feeding the molds.

The arbor used by the gunfounders was an octagonal piece of timber around which straw rope was wound. Loam was pressed into the straw and smoothed off by a strickle board, forming the outside of the cannon (Fig. 92). Next the ornamental pattern marks of the purchaser's coat-of-arms were added and the entire piece dusted with charcoal for parting. More loam having been added, the entire mold was bound with iron bands and baked over a fire, the whole assembly being turned on its arbor until completely dry. Finally, the arbor was removed and the pattern broken away. Usually the breech was molded separately, the whole job being assembled breech down in a pit before the furnace.

Incidentally, in the early 1700s an ordnance law was passed prohibiting the use of elaborate coats-of-arms and decorations,

92 *Applying trunnions and decorative patterns to the gun barrel. At first, most cast cannon and guns bore such ornamental patterns and coats-of-arms, but in the early 1700s a law was passed in England prohibiting their use, after the foundries claimed such ornamentation caused flaws and pinholes. From 1750 on, decorations were engraved.*

because the foundries claimed they caused flaws and pinholes—an early adaptation of cooperation between foundrymen and their customers to the interest of better castings. From 1750 on, all decorations were engraved on guns (Figs. 93, 94).

Even more important than the use of an arbor was the extensive development of separate core manufacture and the use of chaplets to hold the core in the center of the mold. In the writings of Biringuccio there appears a lengthy description of the arbor device for core production and the use of chaplets to hold the core in place. Particularly interesting are his reactions to those foundrymen of his day who professed secret processes having no basis in fact:

"For the first, it is found that only an iron spindle of a suitable thickness is able to support the weight of the clay without bending under the heat of the fire or wobbling as it is turned on its journals or when handled. This should be a *braccio* or more longer than the mould of the gun and worked exactly round, straight and well tapered like a spindle; above all every one of its joints should be well secured and solid. At the upper end a perforated heel should be made, and lower down, exactly at the point corresponding to the length of the mould, are one or two other holes through which the little crossbars can be passed that fasten the core to the outer mould and also serve for constructing the disc.

"Now this iron is to be arranged on two trestles, pivoted so that it may turn truly when rotated. In order that it may not run up and down, make a little iron fork which engages a notch made with a file at the foot, and another at the head.

"After this, you must prepare clay for the core which is well made and tenacious enough not to crack. First take care that there are no pebbles in it and that it is composed of a small amount of cloth clippings and much sifted horse dung and some washed and clean wash ashes so that it may be of a tender and crumbly consistency in order that it may later be more easily extracted from the space inside the gun.

"In the same way, in order that you may be able to take out the iron piece at the same time, first cover it with ashes, and then wrap it all over with a rope of ordinary thickness, leaving a *dito* of the iron uncovered in every *palmo* length. Alternatively, you may cover it all very tightly with hemp tow in the same way after it has been covered with ashes. Or heat your iron considerably and cover it all to the thickness of a good cord with wash ashes or powdered charcoal that has been made into a paste with a little clay water or egg white.

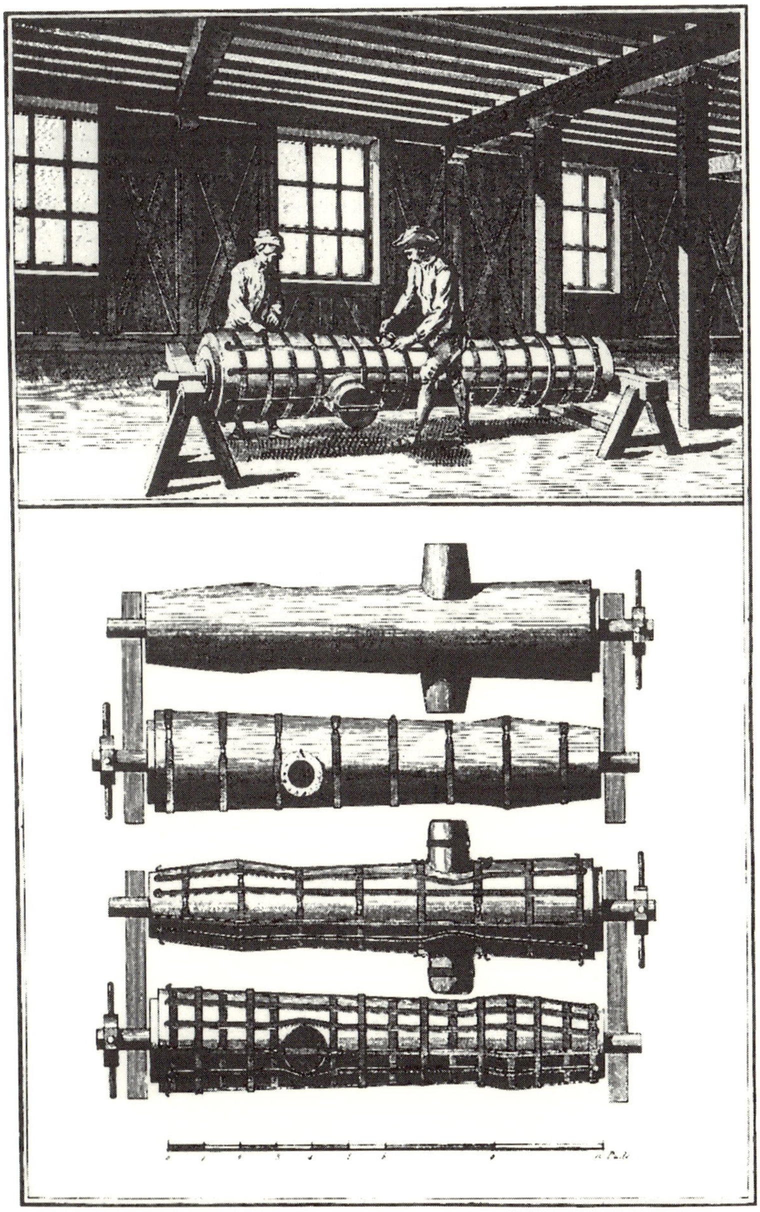

93 *Above: Loam having been applied to the pattern and the parting set on, the gun mold is reinforced with iron bands. Below: Successive stages in the operation of reinforcing.*

"After it has dried take some of your prepared clay, apply a coat of it all over, and bind it with a little hemp tow. Then let this first clay dry very well. Gradually proceed to enlarge it until you are near the thickness that the diameter of the ball is to be, always allowing each coat to dry very well.

"Then take a board of walnut or some other wood that is as long as the core and as much longer as is necessary to reach over its supports, for you are to form it on these. Its thickness is to be such that it holds firm when the clay is put on or such that it will not shake or bend when the core is turned. On one of its sides this is to have a blunt angled cut underneath, and it is to be made as straight as possible with a large plane.

"As I told you above, this is to be fixed firmly with two nails at the points on the trestles where the projecting journals are, but separated from the core by as much as you wish the thickness of the clay to be. Turning the iron spindle, apply more clay on that already put there, and continue to do this evenly until you reach the cutting edge on the board. Then finish it, making it smooth with some soft clay passed through a sieve.

"Having finished and dried it, take it from the trestles and gradually complete with good clay the part lacking at the point and at the foot. Thus when you have completely finished and dried it, cover it all with damp wash ashes. Then when you wish to cast, bake it again in the way that I shall show you in its place.

"There are some masters who reinforce their cores by wrapping them with an iron wire with a spacing of two *dita* between the wires. Some put it on when it is half a *dito* from the correct size of the core. Others do not bother about this reinforcement, saying that the wire causes great annoyance in extracting the core. Without heeding this, I have used either way as I thought best in regard to the quality of the guns.

"There is also among the masters of this art a certain difference of opinion, not yet resolved, in regard to making a part of the foot of the cannon cores which makes a certain difference in the tube, which they call the chamber. Some like it and some do not, and some wish to make it in one way and some in another. Under this veil these men pretend to have a great secret and puff up their reputations by telling lies which deer could not leap over, promising that from their guns not only balls but lightning flashes will issue. But in the end they make only what others have made, and if you ask them what theory moves them, they make only a surly answer."

One might draw a parallel between the sixteenth century, when little was known or written of metallurgy, and the develop-

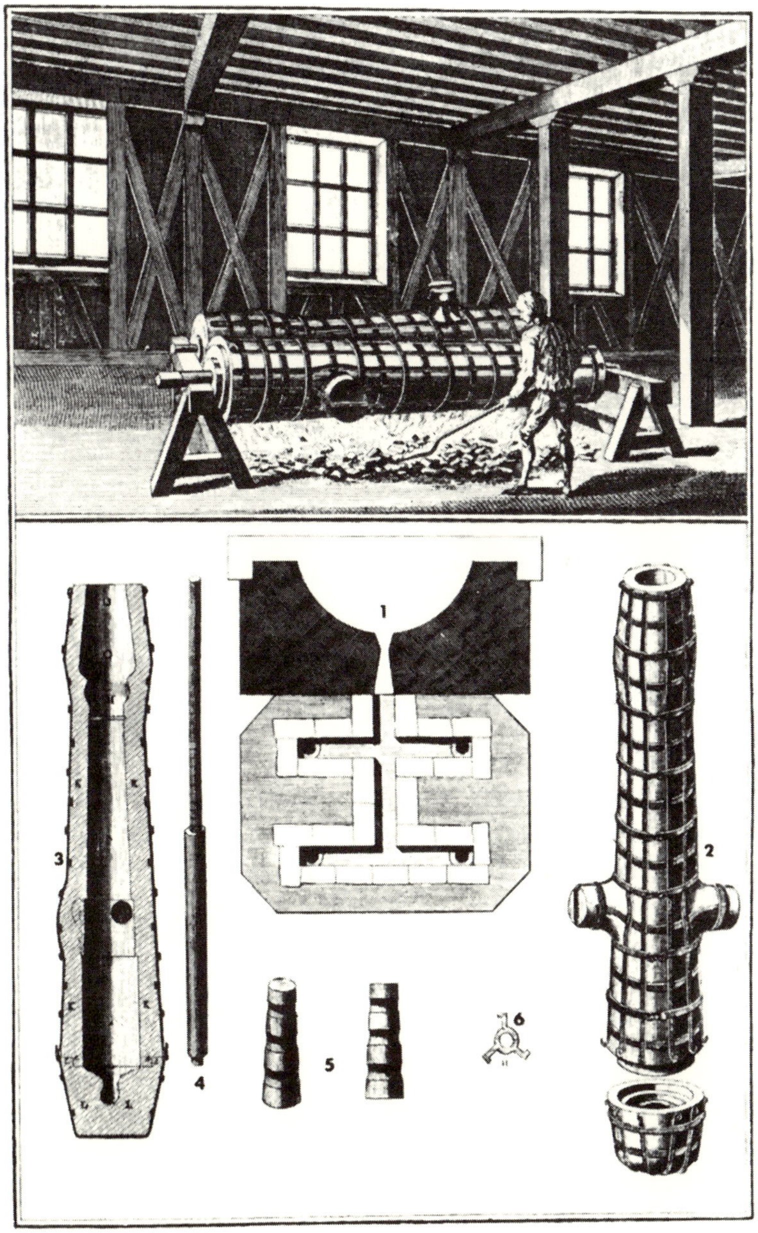

94 *Above: Baking the gun mold over an open fire. Below: (1) Lay out of melting furnace, and pouring arrangements for four guns. (2) Completed mold of gun barrel and breech with arbor removed. (3) Cross-section of gun mold assembled. (4) Core bar. (5) Plugs. (6) Chaplet for core.*

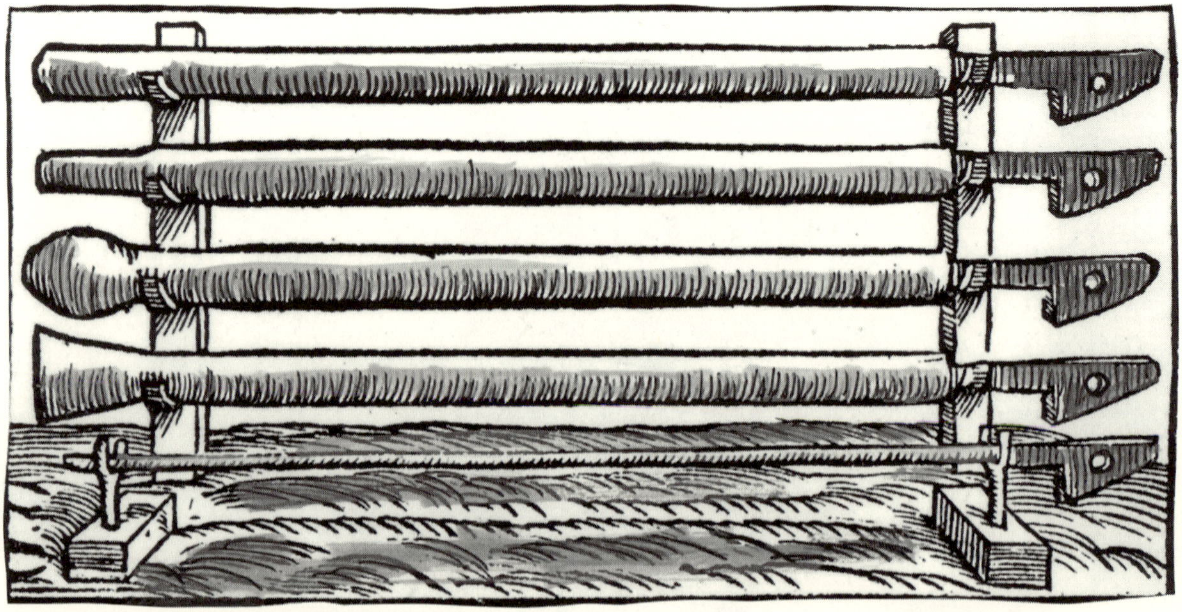

95 *Iron core bar (bottom) and finished cores for gun chambers of various calibers. The use of arbors, cores and chaplets in gun founding was detailed by Biringuccio as early as the middle 1500s.*

ment of modern industry through the willing interchange of information and data. Strange it is to find some men of industry today as in medieval times, still unwilling to admit that the free and open exchange of experience and knowledge is the surest guarantee, if not the only guarantee, against out-moded practices and obsolescent means of production.

"Upon the assembly of the mould," Biringuccio continues, "the core is centered by means of special chaplets. The lower chaplet holds the entire core. The center ring chaplet and the top ring center it in position. The mould is then ready to pour."

At an early date it was realized that a smooth bore was desirable. Thus the boring of guns for smoothness was soon developed because it proved impossible to obtain the proper interior surface by coring alone. The practice of coring guns ceased completely in France in 1713; thereafter all French guns were cast solid and bored. England followed suit in 1775. John Wilkinson, called the "greatest of the early ironmasters," is credited with boring machine improvements that first made accurate machining possible.

Of considerable interest are the melting methods employed in early casting of guns, and the attention given to feeding heads and gates. Biringuccio again is a reliable source for an exact accounting of the actual pouring method. Note the care and thoroughness this master founder, who stressed liberal venting and wide gates in bronze gun casting, gives to all details in the following description of closing up and pouring:

"Now, when this mould has been thus prepared with all the aforesaid attention, when all the junctions are well filled and plastered with moist clay or some other plaster, and when all the gates and vents have been properly formed and fitted with clay plugs which close the places exactly, keep every entrance well plugged either with these plugs, when dry, or with tow. As a further precaution, tie a cloth over it in order to cover it better.

"Understand that if the plugs should become loosened through much handling of the things, it could easily happen that clay pebbles, charcoal or something else, finding no obstacle, would enter the mould. And if not by chance, this could happen by the hand of some wretch who wished to harm you.

96 *Chaplets of various types, as used in Italy in the time of Biringuccio, for supporting cores in the center of gun molds.*

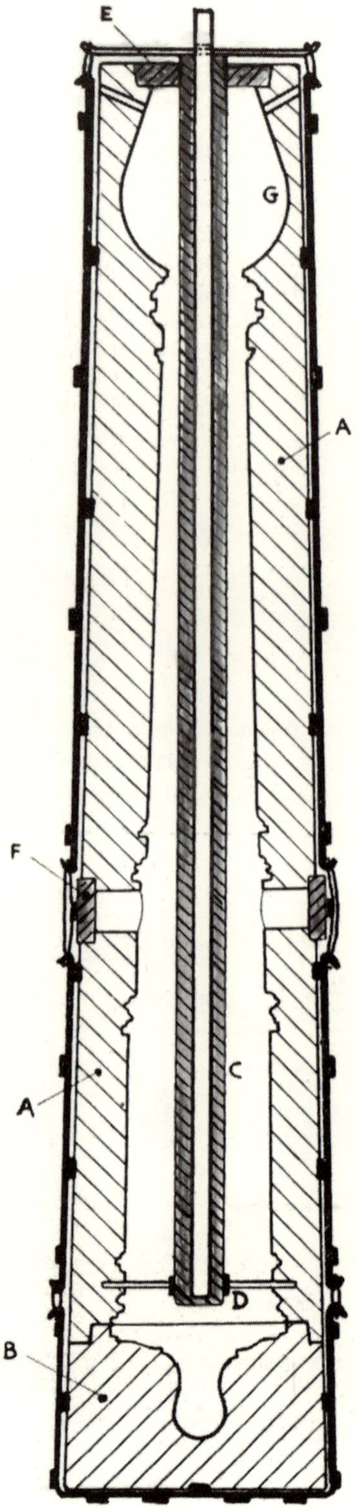

97 *Completed gun mold ready to pour. (A) Mold of gun barrel. (B) Mold of breech. (C) Core. (D) Lower chaplet. (E) Upper chaplet. (F) Trunnions. (G) Feeding head.*

98 *Above: Pouring guns from a reverberatory furnace. Below: (1, 2, 3, 4) Melting tools. (5) Saw for removing gate or head. (6) Completed casting with head still attached.*

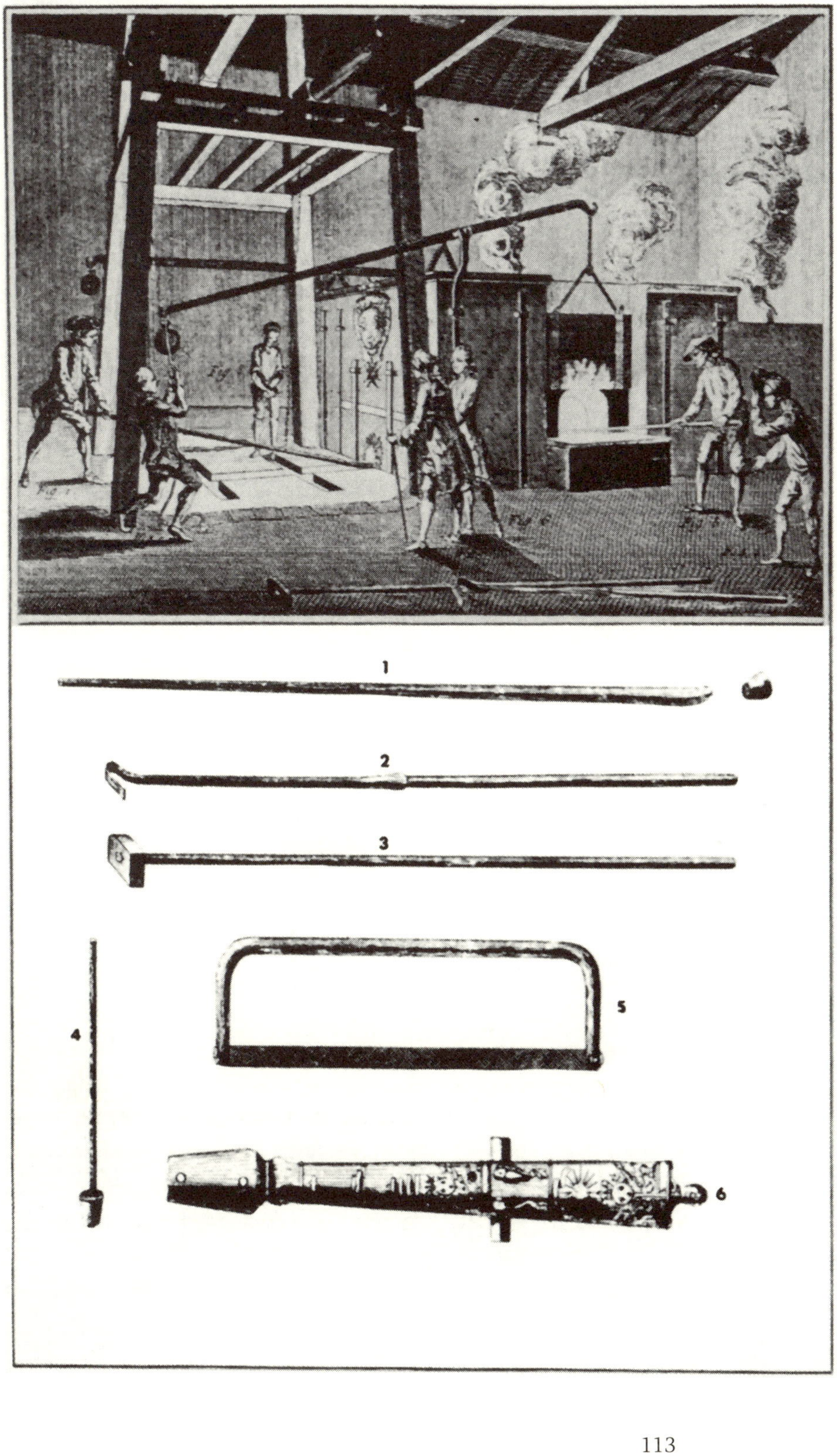

"When you have brought your mould to this point, I advise you to contrive to cast it as soon as possible so that it may not absorb any moisture from the earth, which it is very ready to do. Now arrange this, or these if there are several, in the aforesaid pit opposite the outlets of the bronze, placing their gates so that they may receive the molten bronze through a channel, and taking care that the bottom of the pit is hard so that it may not give way under the weight.

"After you have put your mould in upright, I exhort you above every other precaution to press the earth down well all around it with stamps and to see that you make it hard with the blows. Proceed to put in fine earth in layers of three *dita* or a little more, pressing down with the said stamps without sparing your strength until you have filled the whole pit level with the furnace with the said well-pressed and solid soil. In order that this earth may become better solidified and united together it should be somewhat moist but not wet, for if possible it should become like a mass of stone in this process.

"Having filled up the pit in this way, make a channel of bricks or a clay conduit, a quarter of a *braccio* in both width and height, between the mould and the outlet for the bronze in the furnace. Cover it with ashes and then bake it very well with charcoal or wood, so that when you see that the bronze is melted and ready and you take it out through this conduit to bring it to the gate of the mould it may feel no moisture or coldness.

"Having done this and having also baked the bowl, when you see that the bronze has arrived at the good and true readiness of its fusion, clean the channel and the bowl of the gate of every tiny thing and, by striking with an iron rod in the hole that was arranged as an outlet of your furnace, cause the bronze to run out like an oil."

Before leaving guns and ordnance, a description of the cast cannon ball is in order. As stated, concrete balls were used until 1568, after which the iron ball came into being (Figs. 99, 100, 101). Many methods were employed in casting these balls, including sand molds as well as complete permanent molds. One of the most unusual methods is described in Louis de Tousand's *Elements of Artillery*, in an interesting combination of casting and forging:

"Cannon balls are cast of iron, the metal running into iron moulds divided into two parts, which are denominated shells, fitting each other exactly. The superior part sets on the other which is laying on its flat side; it should be sufficiently heavy to prevent the liquid metal from raising it. The vent or aperture of the set through which the liquid metal is to be introduced, to fill up the

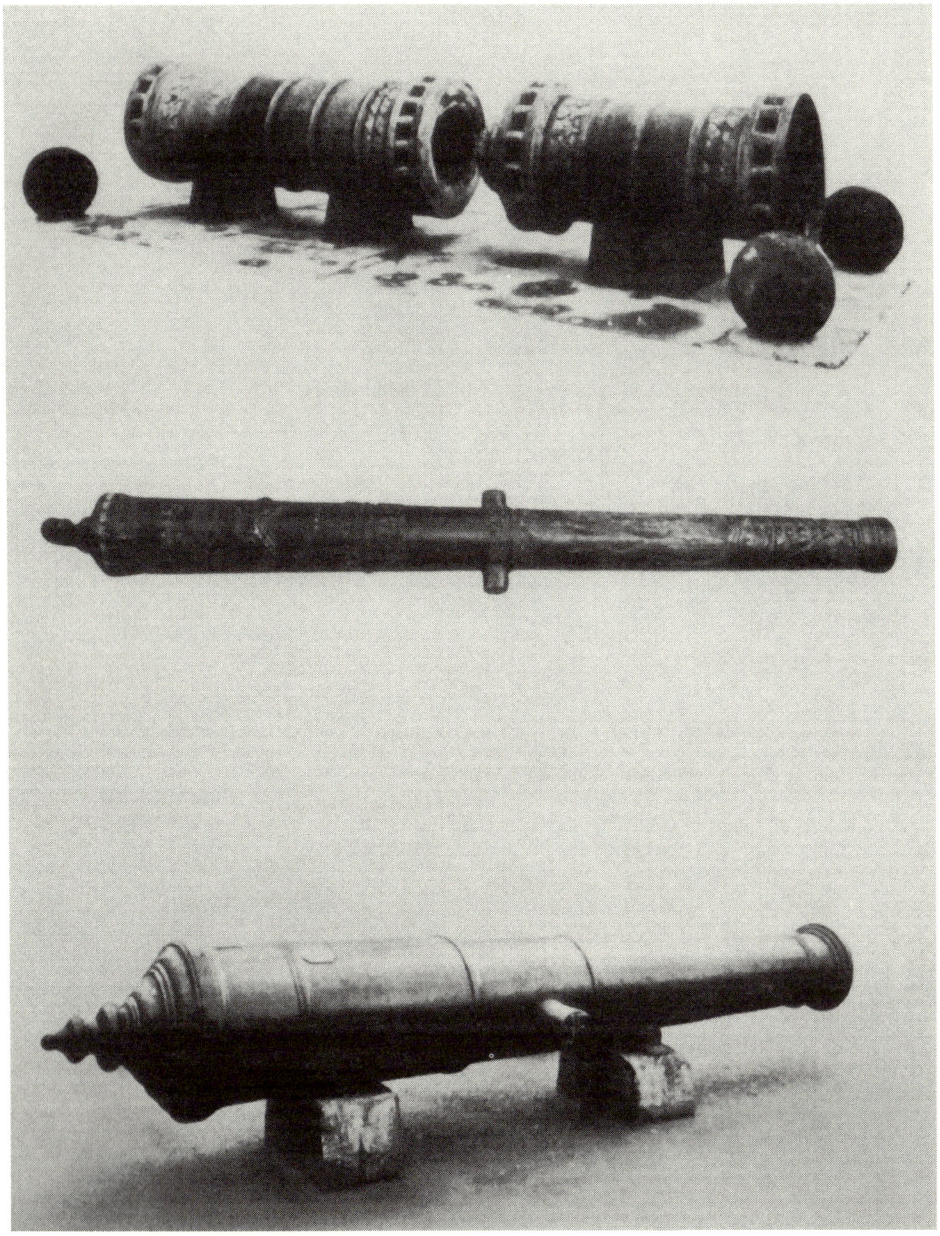

99 *Some early types of cannon; above, the Dardenelles gun of the bombard type, middle, Spanish serpentine of cast bronze and, below, English cast iron cannon.*

100 *Production of cannon balls, using a small hand-blown bellows furnace. Iron shot produced by this method usually involved a combination of casting and forging.*

mould, is made in this superior part. The liquid metal should be conveyed to the mould in a small stream.

"When taken from the moulds, the balls are imperfect and their surfaces seldom smooth. In order to smooth these unevennesses which might scratch the bore of the cannon, the balls are heated somewhat more than cherry red in air furnaces, out of which they are withdrawn with a strong pair of iron tongs and placed on an anvil hollowed concave, one quarter of the shot's diameter deep; on this anvil they are hammered with heavy hammers of the same concave depth.

"These hammers are put in motion by means of water or other machinery, and in the interval of the strokes, a workman holding the ball with the iron tongs turns it on all sides until it becomes perfectly round; from one hundred to one hundred and thirty strokes are commonly sufficient for this operation."

For many years after the invention of cast cannon it was not inconsistent to find bells and cannon alternating in the founders' pits. As nations fluctuated between war and peace in medieval times, bells were recast into cannon, which in turn were remelted and molded bell-shaped once more. Thus we find described in the *Annals of Methodism* an eighteenth-century gun foundry, the "Old Foundry" at Moorfields, which became a place of worship, the church of John Wesley in London.

Description of bells and guns practically completes the introduction to the era leading up to modern foundry practice. However, two circumstances associated with the thirteenth and sixteenth centuries should first be mentioned, since both are essential to any history of the foundry industry. One concerns a development of tremendous importance although seldom associated with the foundry industry today; the other concerns a man who was both a great artist and a founder—Benvenuto Cellini.

In the middle of the thirteenth century came the invention of movable and cast lead type. No single invention since the melting of metal has been so vital to civilization and its progress as the development of cast type and the printing press (Fig. 102). Actually,

101 *Permanent cannon-ball mold (note dowel holes) and the finished cast ball obtained from such a mold.*

the invention of cast type far overshadows other inventions in importance, for without it the vast printing and communications industry as we know it today would not have been possible.

Type founders are merely foundrymen working with cast alloys and permanent molds; yet to the writer's knowledge the literature on cast metals does not contain a single mention of any connection between this admittedly great development and metal casting in general. This well may be an indication of how completely the world has come to accept casting as a basic process, to the point that a casting is viewed not merely as a cast product but rather as an object of great industrial usefulness.

Although the emphasis in this chapter has been on cannon and bells, it should be borne in mind that the close connection between the foundry and art also survived the Middle Ages (Fig. 103). Best evidence that the foundry industry's artistic traditions—sculptures in bronze, the molding of precious metal jewelry, and classical art expressed in cast metals—did not die, is found in the masterful works of Benvenuto Cellini.

102 *Woodcut of early type founder casting lead type, the beginning of the printing press. Although the literature on cast metals contains few, if any, references to type casting, type founders are merely foundrymen working with cast lead alloys and permanent molds.*

103 *Bronze cast door at San Giovanni, Florence, Italy, designed by Ghiberti. Cast in the fifteenth century, panels represent scenes from the Old Testament. The beauty of the panels caused Michelangelo to name the door "Porta del Paradiso."*

104 *Exquisite gold salt cellar cast by Benvenuto Cellini in Florence, Italy, about 1540, a famous product of the founder's art.*

In this famous Italian artist, who contributed so much to foundry progress in the sixteenth century, were combined the talents of the sculptor, the silversmith and the foundryman. His gold salt cellar (Fig. 104) and his bronze Perseus, cast in Florence about 1540, are both world famous and a tribute not only to his artistry but also to his foundry ability. A child of the Renaissance, Cellini has been credited with the rediscovery of lost-wax molding; but this claim seems to be erroneous, since Cellini actually did most of his work in loam, as did all founders of his day. Wax probably was used for parts of the pattern; but loam molding was then in full use, and he undoubtedly employed the most modern methods of his time.

One story concerning Cellini may strike a sympathetic note with all foundrymen and pertains to his casting of Perseus (Fig. 105). Cellini described in detail, in his *Autobiography*, his preparations to cast: how he completed the mold, vented it, drew out the wax and lowered the mold into the pit for pouring; how he started his furnace and charged pigs of copper and bronze scrap. From these exertions he became violently ill with fever and retired to his home, leaving his "faithful" molders in charge.

At home he awakened from a terrible dream and with a premonition that something was wrong, rushed to his foundry. In his absence, and due to neglect, the roof of the foundry had caught fire,

105 *Cellini's world renowned bronze Perseus and Medusa head reputedly cast in 1540, and a splendid tribute to the maker's artistry and foundry ability Cellini worked chiefly in loam molding, and Perseus was cast in such a mold, in the face of amazing (even amusing) difficulties. (See text.)*

and to make matters worse a high wind had blown in all the windows of his little shop. While he and his helpers put out the fire and hung rugs at the windows, his fire went down in the furnace and the melt began to "freeze."

Cursing his melters for ignoramuses, Cellini built up the fire and threw in his remaining pigs of metal. Needing still more metal, he threw in all his pewter dishes and then poured the mold. As soon as he had finished, the furnace blew up completely; but the Perseus was finished, and the result was a world-renowned casting. Perhaps there is some solace here for all modern foundrymen, who have a great many troubles all at one time but who generally produce good castings in spite of all obstacles.

CHAPTER 6

Sand Castings of Direct Blast Furnace Iron

As Europe recovered slowly from the shattering of Roman civilization, trade and commerce gradually revived and the rise of a "free" industry brought into being the craft guilds. During the latter years of the Middle Ages and into the eighteenth century, they played a prominent part in commerce and industry, including metal casting.

These guilds exercised complete control over all skilled workers. Not the least to be affected were the metal-casting establishments, including the non-ferrous founders and an increasing number of foundrymen who were turning to direct-furnace iron casting. These guilds exerted a great influence on foundry operations, for each trade or craft, including metalworking and founding, had its own guild for the purpose of improving workmanship and controlling quantity and prices.

In some ways the guilds accomplished a great deal of good. They established the principle that to do good work is an honor, and they set high standards of workmanship, quality and honesty. Shoddy work or short measure was punished severely by lopping off ears or hands. The guilds protected their members in times of sickness or death, and prescribed wages, hours and work conditions; but they also artificially forced the distribution of work.

For instance, no master molder could have more than four journeymen and two apprentices. He could not compete with a rival or take work from a customer who had mistreated a fellow guild member. Such rules, carried to the extreme, together with the stagnation that sometimes is associated with monopolistic power, could not last, and the guilds finally deteriorated.

106 *Woodcut of old charcoal blast furnace in Wales, probably built about 1600* A.D.

However, into the atmosphere of guild control was projected one of the most critical rediscoveries of European industry—the melting and casting of iron. To the lasting credit of guild workmanship must be laid much of the progress in iron founding which ultimately affected the entire metal-casting industry.

The manufacture of pig iron from the blast furnace began in the early part of the fourteenth century, after having been discovered much earlier in the Orient and used again in Greece to some extent. Now a new era of casting ferrous metals was developing in Europe. It should be carefully noted and continually recalled that all castings in iron from about 1300 to 1700 A.D. were made from direct-blast-furnace metal. The production of iron by cupola or other remelt methods, as we know it today and involving close chemical and metallurgical control, was unknown. Some remelt existed, it is true, but the stove plate, firebacks and iron cannon made at this time were for the most part cast from direct-furnace metal.

The field was limited by the high percentage of carbon and resulting weakness in metallurgical structure. Not until the advent of coke and the steam engine, providing a reliable supply of fuel and air for the cupola, did iron founding come into its own.

107 *Charcoal blast furnace looking into the forearch, from which iron was drawn off into pigs. All iron castings from 1300 to 1700 A.D. were of direct furnace metal, produced in similar blast furnaces.*

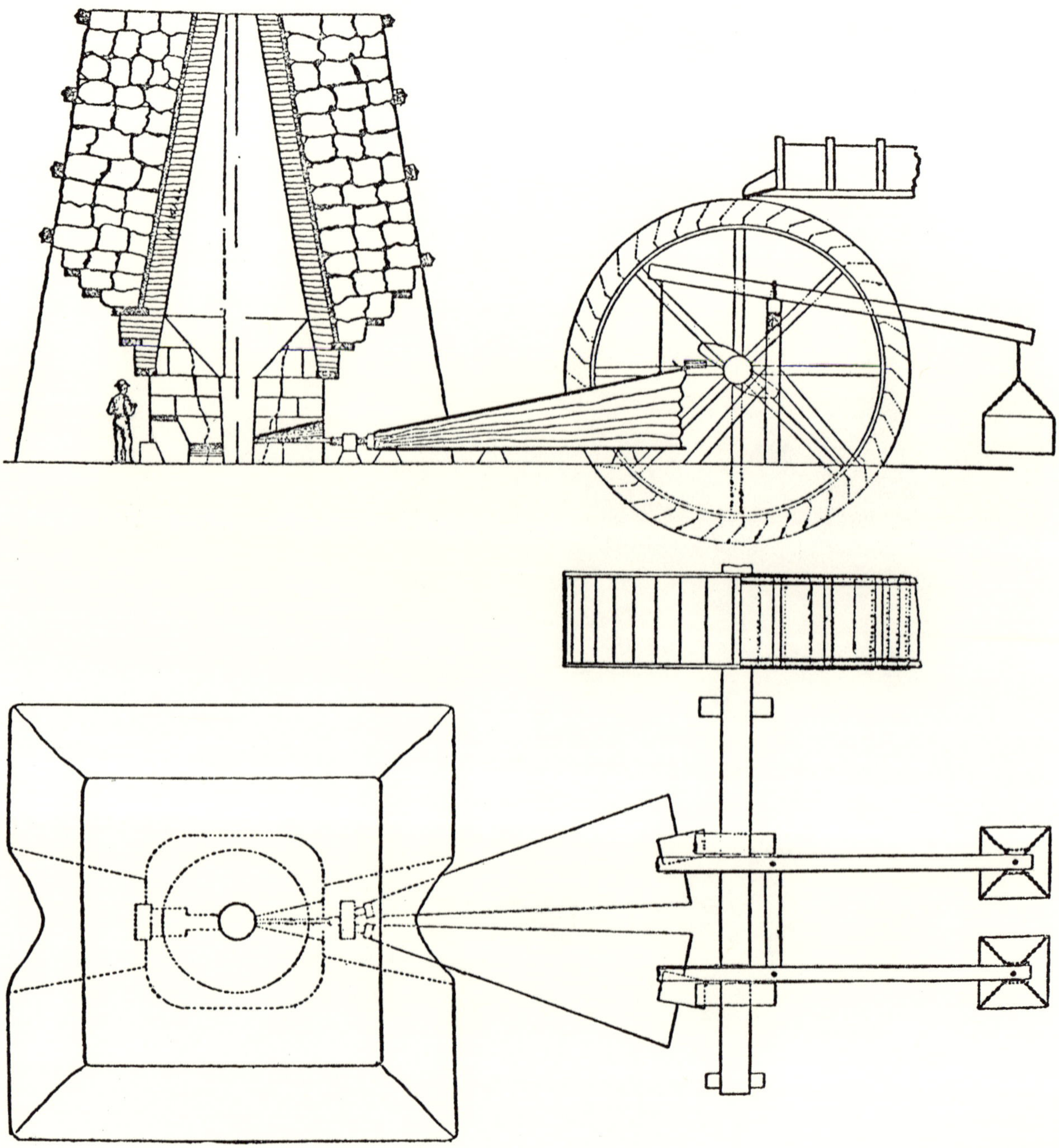

108 *Plan of a blast furnace actuated with bellows, driven by an overshot waterwheel, water being delivered from a flume. Metal is melted in a crucible set into the hearth and flows on removal of a dam at hearth level.*

109 *Diorama of early blast furnace, showing formation of pigs in front of furnace, charging bridge on top of furnace, and waterwheel-driven bellows at right.*

Before this historic invention, however, the methods and castings developed are of greatest importance and should be analyzed in the light of their general use today and their effect on modern practices. The charcoal blast furnace has already been described (Figs. 106, 107), with its development from a mere hole in the side of a hill to operation by means of a waterwheel-powered blast.

Generally speaking, ore was obtained from nearby bogs and calcined for two or three days, thus preparing it for reduction. The ore was charged into the top of the furnace, together with alternate layers of charcoal and fluxing agent in various forms. As slag appeared it was hooked out, until finally a bloom of iron was obtained and hammered into shape. By utilizing the water-driven bellows and increasing the height of the furnace shell, molten iron was obtained (Fig. 108). This pig iron was run out through sand channels and allowed to solidify into small pieces (Fig. 109).

Although no exact date for the introduction of European pig iron has been established, it probably was produced in reasonable quantities in Italy and Germany by 1400, and in France and England by 1500. These are the dates most generally accepted.

Castings of furnace metal have been discovered in Germany dating back to 1313. Cast pipe was produced in 1455. Stove plates and firebacks have been traced back to Alsace and the year 1490. Grave slabs and markers of iron were used in England in 1450.

Unquestionably, the art of sand casting came into Europe from China, where iron castings were poured in sand molds as early as the sixth century B.C. Since the trade routes to China were reopened in the sixteenth century, it is not difficult to understand how the Europeans developed and perfected sand casting on their own initiative in the short space of time that elapsed between the melting and the casting of iron.

Furthermore, it is evident that European metal casters were familiar with sand molds even before the coming of iron. Leonardo da Vinci (1452-1519) set down the advise that "If you wish to make casts rapidly and simply, make them of a box of river sand moistened with vinegar." Biringuccio, contemporary with da Vinci, also speaks of sand molds in connection with non-ferrous casting, and specifies the following procedure:

"Contrary to the natural order of art, it has been discovered how to cast in moist earth in order to avoid labor and expense. This is truly a thing that many desire and few practice because it is not as smooth a way nor as easy to effect as it apparently seems.

"To do this, one part of fine-grained yellow tuff is taken, or some very fine river gravel that has been very well washed and baked in a furnace. Then a third part of young ram's ashes is taken and a twelfth part of the whole of old flour that is finely sifted. By pounding, all these are incorporated and mixed together well.

"Then take urine or wine and moisten it, and mould whatever you wish with the moistened material in frames or wooden boxes. Having taken out your reliefs, make the gates and vents, if you did not make them at the same time that you moulded the object. Then cover them with soot as usual, with the smoke of turpentine or a tallow candle. Then the moulds that you made are again fitted together and are cast at your pleasure when your metal is melted." (Fig. 110)

The key to the assertion that sand molding was known in Europe as early as the sixteenth century is seen in Biringuccio's

110 *Green sand mold and core for casting a small bell in the thirteenth century.*

statement that sand molding did not give as smooth a surface as loam molding. Surface appearance was of primary importance to those early founders—and still is today!

However, as higher metal temperatures were obtained, sand was added to the loam and the desired results achieved. Thus the combining of two known molding procedures so as to obtain a new process came as a gradual adjustment to conditions rather than as some sudden new discovery.

Unquestionably, the development of iron founding received its first major impetus through public demand for firebacks and stoves. In our modern day it is difficult to appreciate an adobe without so much as stove heat, yet the stove is only about five hundred years old and did not come into being until a foundryman was able to provide fire-resistant materials for its making.

First came the fireback. This useful casting was placed at the rear of the fireplace, thus making it no longer necessary to replace refractory backwalls at frequent intervals. Iron castings for firebacks were turned out by the hundreds of thousands early in the sixteenth century.

111 *Open roller mold, in some respects the first crude type of molding machine employing sand. (A) Roller for compressing sand. (B) Rail on which roller moves. (C) Vent holes. (D) Sand, cinder and charcoal bed.*

112 *Open roller mold (same type as in Fig. 111), with woodcut dies in place for molding. (A) Runner from main reservoir of metal. (B) Pattern dies in place.*

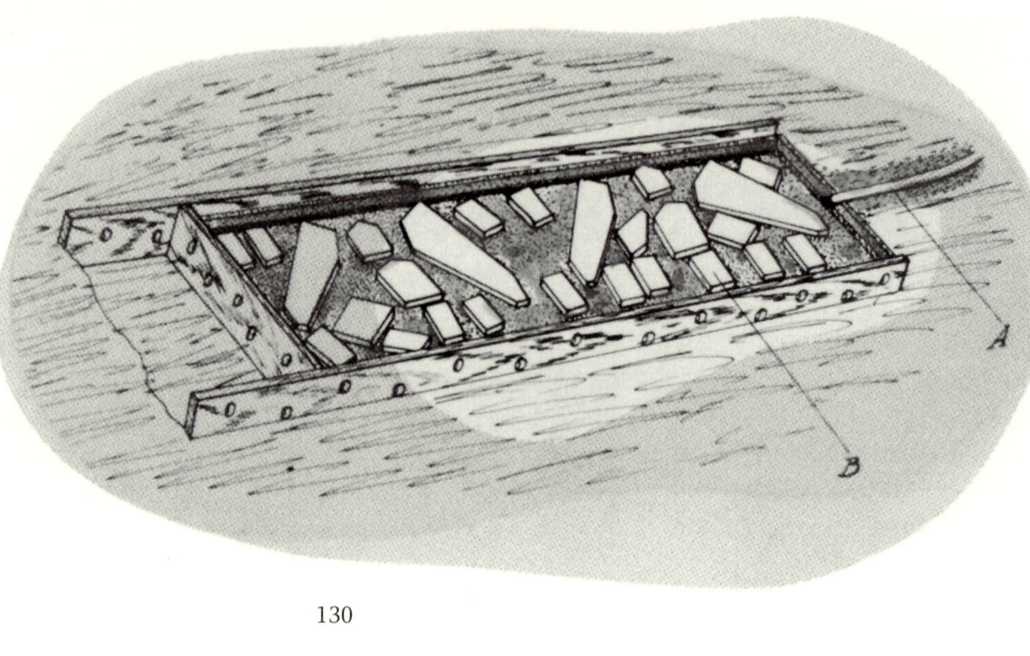

113 *Fireback casting obtained from mold such as shown in Figs. 111, 112. Note uneven definition of figures, due to varying pressure exerted by the mold roller.*

After the fireback came the stove, first for the wealthy and then, as the result of casting economies, for every home, however poor. It should be noted at this point that these products had brought about two important changes in sand castings, both of far-reaching effect. The first change occurred in the early molding equipment used for open sand molding. The second change introduced that most necessary trade, patternmaking.

Prior to the use of patterns it was essential, of course, that the mold first be prepared so as to receive the pattern decoration and then the metal. For this a most interesting device was used which constituted, in part, the first molding machine employing sand.

Two wooden rails were placed parallel on the foundry floor with the tops slightly above floor level and the bottoms some six inches below. One end was blocked off, while the other end led to the furnace by means of a channel. Between these rails a bed of cinders was laid, upon which was sifted a level of sand and then a level of charcoal dust.

The molder then dropped on to the rails a heavy roller having a diameter of three or four inches, and by rolling this back and forth produced a smooth and level pit with the sides and ends protected by several inches of sand (Fig. 111). Into this bed the molder then pressed individual woodcuts or dies to suit his artistry. Frequently he used three or four woodcuts in some regular arrangement, and at other times he pressed them in with no apparent object save to relieve the surface (Fig. 112).

114 *A foundryman's own fireback design. Lower left shows the blast furnace. Upper left shows coat of arms, including hammer, weights, pliers and slick. In the center is the foundryman himself, with ladles and weights between his feet. Lower right shows reproduction of a fireback bearing the master's initials. Center right shows a pig, symbolic of a foundryman. And, in upper right, are several jugs of wine—even then, it would seem, somehow symbolic of a trade-developed thirst.*

115 *Fireback showing use of pattern dies. The die for "1632" evidently was mounted on a block, while the initials were hand stamped. A rope was used to form the border design.*

116 *Ornamental box stove made of castings and fitted into a wall so as to be fired from the fireplace in the adjoining room.*

After these dies had been removed, the molder patched and finished the mold and either opened a channel to the furnace or poured the mold by means of a ladle. As soon as the metal was set, he piled on weights to prevent warping. When cool, the casting was pulled out (Fig. 113), and the sand in the mold was brushed up and inspected for shot and scrap; then the molder was ready for the next pouring.

Thus one man performed all the necessary tasks, from molding and pouring to the shakeout process. The advantage of this roller-and-rail method of open sand molding (Fig. 114), lay in its flexibility of widths and lengths, and in the fact that all work could be accomplished at a point easily accessible for gravity pouring. Such was open sand molding in the fifteenth-century iron foundries (Fig. 115).

Prior to this time little use had been made of patterns, other than destructible wax patterns for non-ferrous work. Germany,

117 *Stove-plate casting showing (above head of man) where the loose pattern shifted in molding, causing the die bed to become plainly visible in the metal.*

where the box stove was developed, was also the home of the most excellent woodcuts (Fig. 116), and it was only natural that founders casting a long, plain piece of iron would desire to ornament it (Fig. 117). Individual woodcuts were used for this purpose and pressed into the sand one at a time. This accounts for the occasional appearance of the die face in firebacks and stoves, resulting from a bad impression (Fig. 118).

The craftsmanship of individual molders also appears in the various trademarks they affected. A roll of tobacco, a clenched fist or an open hand are to be seen frequently on old stoves—all trademarks used by molders in an effort to identify and adorn their work.

This sort of hit-or-miss mark died out as greater numbers of stoves and firebacks were produced, when ornamentation then became a requirement. In an effort to obtain uniformity in quantity production, the molder mounted individual woodcuts or dies by

118 *Stove-plate casting for which the same dies apparently were used twice (left and right). The molder who molded this casting seems to have been rather heavy handed, for all die edges show plainly.*

119 *Oven plate cast from a pattern having all dies mounted on a board. Some shifting of the triangular ornamental dies at top can be noted.*

screwing them into a board (Fig. 119) so that all his dies would have equal depth and detail. After pressing the whole board, he would unscrew the dies and return them to the mold one by one in order to obtain a clearer impression of each.

Eventually the entire pattern was made in one piece and bolted to cross-boards to facilitate pressing into the sand, as seen in Figs. 120, 121. Ultimately, a metal pattern was developed when it became evident that abrasion wore down the sharpness of detail in the wooden pattern (Figs. 122, 123). Metal patterns were not used in great numbers until the year 1800, although complete wood patterns go back as far as 1700. This date, however, applies only to patterns employed in sand molding (Figs. 124, 125, 126).

Johannes Gutenberg used metal molds for casting pewter pots and tableware in the latter part of the fifteenth century, but his process is more applicable to permanent molding than to pattern work. It seems apparent that an unusually advanced type of pattern

120 *Face of complete wood-carved pattern. (Compare Fig. 121.)*

121 *Rear of complete wood pattern illustrated in Fig. 120, showing that the pattern was made of two wood boards bolted to a crosspiece to facilitate pressing into the sand. (Note: To the left of normal joining, pattern has apparently split.)*

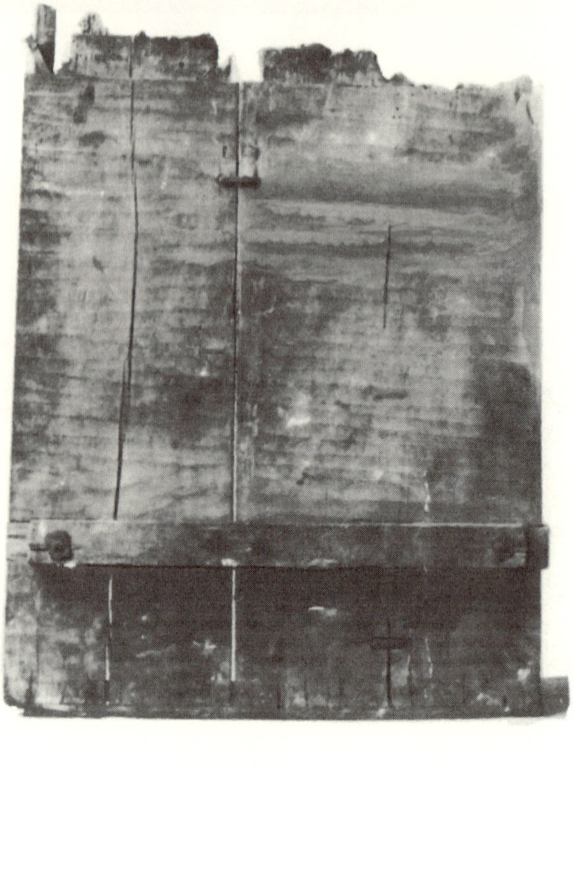

122 *Complete one-piece oak pattern made in Germany in the early 1800s.*

for non-ferrous work was in use in Italy in the sixteenth century, although it astounded even Biringuccio, who described its use in his usual concise manner, as follows:

"There were eight masters near others in a room who did nothing but mould *in lute* and form an infinite number of moulds of all those little objects of everyday use which can be made from brass by casting. They made these in such a fine way that I cannot fail to tell you about it.

"These masters took the number of patterns of all the things that they had decided to mould, that is, harness buckles, cups, belt buckles, all kinds of chain links, bells, thimbles, window fastenings, and other similar things. One day they made all of one kind and the next day all of another, and thus they proceeded, changing the pattern every day; and having finished one that they had to mould, they began again at the beginning with another, continuing always with this easy method and process of moulding to accomplish much work.

123 *Early metal pattern dating to about 1800. Some metal patterns of this period show a repetition of design, indicating the use of a master pattern. Metal patterns were not used extensively prior to 1800.*

"They took a large quantity of *lute* mixed with cloth clippings or cane seed and, when they had beaten the quantity they wished to use until it was somewhat hard, they spread it out about half a *dito* thick or less on a little board which was a *palmo* in length and somewhat wider than the patterns. After having spread it out well, they dusted it with fine charcoal, and in it they moulded their patterns all attached to the gate, with vents, openings, and all the parts which are needed to make a complete mould.

"Some of these patterns were of tin and some of brass, accurately made, filed and well finished, for if once the mould was well made the objects had to come out the same way."

This unusual use of patterns, complete with vents and gates, was never again mentioned by Biringuccio; hence, it can only be

124 *Woodcut used by an early pattern-maker for carving his one-piece pattern. Origin apparently German. (Compare Fig. 125.)*

125 *The casting that resulted from copying the woodcut illustrated in Fig. 124. Evidently the patternmaker was more skilled than the original artist.*

126 *The ultimate in ornamentation of stove plates. This excellent casting, made about 1550, reveals quality workmanship all the way from the pattern to the finished product.*

assumed that he did not follow up the idea. However, the woodcut dies of the German patternmakers seem to be the true ancestors of the modern pattern.

Germany followed the art of woodcut patternmaking for many centuries. Between 1776 and 1825 the famous Berlin cast iron was being produced under royal patronage. Beautiful examples of this cast iron are now, fortunately, the property of an American foundry concern, The American Cast Iron Pipe Co. of Birmingham, Alabama, where they have been preserved as part of the exquisite Lamprecht Collection (Figs. 127-130).

127 *Ornamental stove door made of Berlin cast iron, a metal developed between 1776 and 1825. Many fine examples of this iron are in existence today.*

128 *Exquisite animal statuettes cast in Berlin cast iron by Kurek of the Magdesprung foundry, Germany.*

129 *Intricate iron jewelry of Berlin cast iron, so delicate that the castings are wired together. For the famous Iron Cross, Germany's highest military decoration, the metal chosen was Berlin cast iron, indicating the esteem in which the metal was held.*

The man who developed Berlin cast iron to its highest point—Stilarsky, a German combination of sculptor and foundryman—also was responsible for invention of the split metal pattern so necessary to modern sand casting. He also developed facing and backing sand and realized the necessity for hot metal in thin-section castings.

With the development of Berlin cast iron, sand castings and patternmaking were established on a firm foundation. Products which the public wanted, and methods by which they could be produced economically and in quantities, became possible. Because of their durability and economy, castings were making it possible for the average man to enjoy some of the comforts of life—household articles, jewelry, art work, tools—at relatively low cost.

130 *Filigree fruit bowls, sandwich trays and plates of Berlin cast iron. All examples of Berlin cast iron shown (Figs. 127-130) are from the Lamprecht Collection, the property of American Cast Iron Pipe Co., Birmingham, Alabama.*

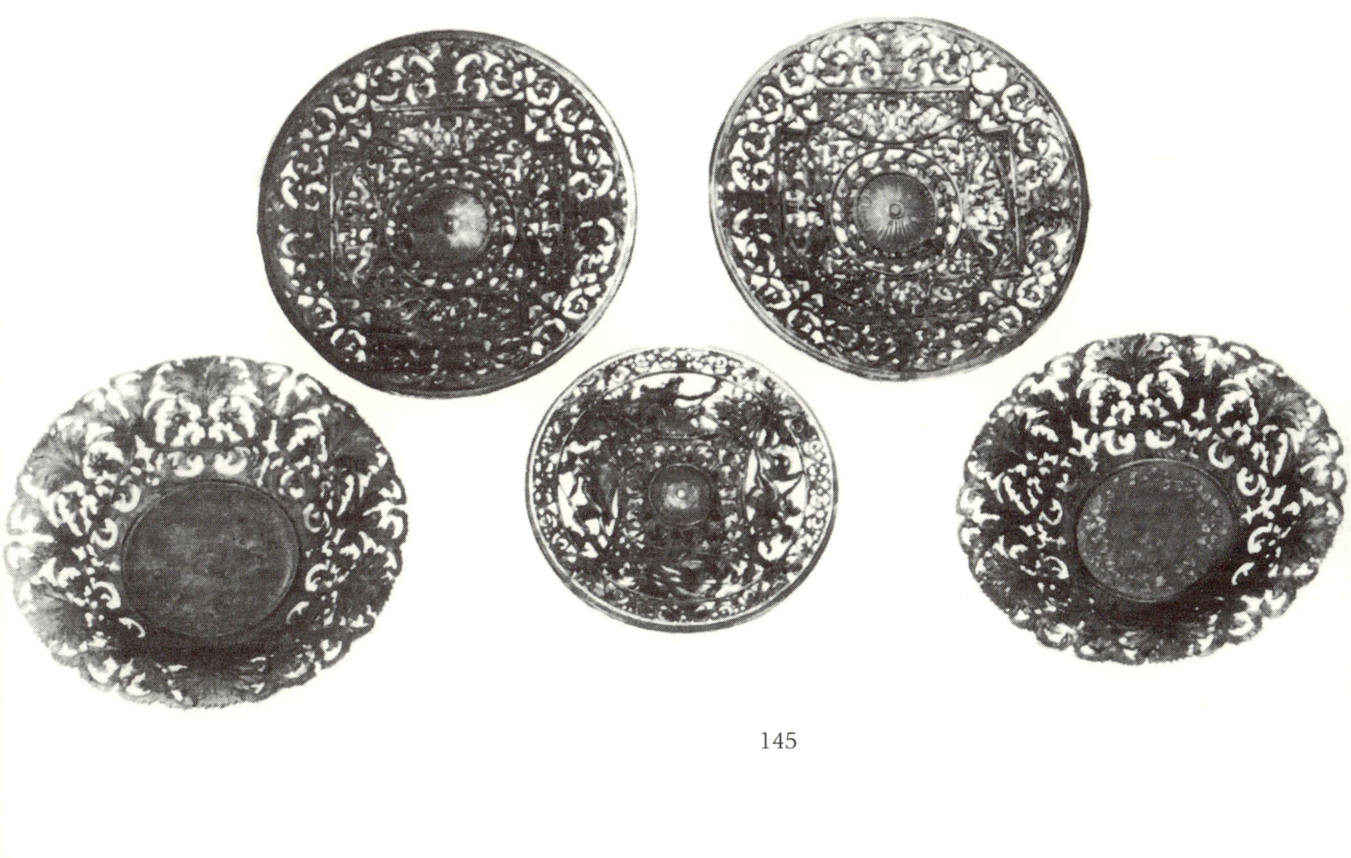

No other industry had proved so beneficial to the "little man," even though the development of castings still was in its infancy in point of service to humanity. Up until 1700, iron castings were of the direct-metal type and, as such, had great limitations. Now iron prepared to enter a newer and greater era of development in which remelt, alloying and refining would make it possible to offer even finer products for the pleasure, comfort and safety of every community.

In the production of iron castings the foundry industry again demonstrated its importance to mankind. Already ages old, it was still young and growing, widening its service, proving itself a truly basic industry as "a necessary means to very many ends."

CHAPTER 7

The Early Iron Foundry

THE SIXTEENTH and seventeenth centuries in Europe saw tremendous strides made in the use of the relatively new cast metal, iron. Non-ferrous metals also enjoyed equal growth; but the greatest attention was now centered on iron, which previously man had apparently considered the least important of his metal discoveries.

Iron had not found immediate and widespread acceptance, and frequently was referred to as the "base" or "corrupted" metal, implying greater favor for the more attractive bronze or brass. As late as the 1780s, when cast iron was replacing timber for construction purposes, the engineer John Smeaton, who had improved the blast furnace and employed cast iron successfully in many fields, had such prejudice to combat. Few good words had been said for iron, except by one hopeful (and badly advised) chemist, who prescribed iron to be taken *internally*, thus:

"It [iron] is harmless in effects on the human constitution, when taken in the raw state or in the form of rust, unless in immoderate quantity; a quality which few other metals possess."

Cast iron had to demonstrate its serviceability in spite of prejudice, and so its earliest applications were not exacting—tomb slabs, firebacks, weights, anvils, cannon balls, stamping irons and the like. Here the material quickly proved its usefulness and economy.

As costs and prices mounted in Europe, iron applications multiplied and blast furnaces spread, particularly on the Continent. While production was small in comparison with today's tonnages, it was large enough to strip great areas of Europe's forests. Before the close of the 1500s, cannon and shot were taking substantial tonnages, and domestic needs included hearth equipment, household and industrial utensils, mortars, pipe, stoves and many other items.

Few men of those times fully realized the power inherent in this, the most common of all metals. Yet this ignorance of the sixteenth century is somewhat akin to the indifference of the general public today, who take iron and its casting so for granted that a proper appreciation of its utility is far from general. When it is considered that our modern world would not be possible either structurally or economically without the widespread use of cast iron, the general underestimation of the present day becomes far less understandable than the attitude of man of the 1500s, who had little or no experience to guide him.

It is not intended here to hold a brief for any one cast metal. All have their virtues and their place, and there is room for all in this mechanical world of ours. However, the cause of cast iron, which includes the majority of cast tonnages poured, has not received proper recognition from either the average man or the engineer. There exists a failure to appreciate not only the structural possibilities of iron but, even more important, its economic advantages.

Its economy, a contributing factor to this lack of recognition, has made possible many products which the average person now accepts as a matter of course, but which would have been considered miraculous only a few short centuries ago. Iron and the ability to cast it made it possible for Watt to create the steam engine, and for others to develop the internal combustion engine, so familiar in our modern automobiles; the pipe and conduit delivering oil, gas and water reliably day after day; the electric dynamos at the basis of our vast electrical systems; and the thousands of different types of agricultural and industrial equipment. True, all of these might possibly have been developed in other metals, or by some process other than casting; but the relatively low production cost of cast iron in many intricate shapes has materially aided the widespread use of these products.

To some it may appear that the day of the iron foundry is ended; actually, it is only beginning.

Up to the time of the seventeenth century, cost had somewhat restricted iron casting to items for war and for the wealthy. Development of the stove, which brought with it comfort and health for a great majority of people, was due entirely to the art of casting iron. Thus far the use of the stove has been dealt with mainly in connection with patternmaking and sand molding; but

131 *Hollow-bored logs were used as water pipe in the fourteenth century, prior to the advent of cast-iron pipe.*

unquestionably this unspectacular, though useful, cast product had much to do with improving the living conditions of the average man and broadening his field of enterprise. Before 1800, stovemaking had become one of the "most extensive, superb and profitable manufactures" in Europe and America.

Another equally important cast-iron achievement of the fifteenth century was the development of cast-iron water pipe. Cast-iron pipe did not become possible until the advent of molten iron and the ability to cast it. We will never know how many thousands of lives were saved from epidemics by the use of cast-iron pipe, but we do know that its use has added immeasurably to the comfort and convenience of the average citizen.

Today cast-iron pipe is such a necessity, and so common, that it is taken almost completely for granted. Millions of miles of such pipe are underground in the United States alone. Few realize how greatly it has contributed to civilization by making it possible to supply towns and cities with pure water and gas, and to dispose of sewage quickly and safely.

Prior to the development of cast pipe and iron, man had tried for years to pipe water to his residences and workshops—cheaply, conveniently and surely. In the early days of the fourteenth century, water was delivered to homes through open ditches, which were subject to every possible pollution. The Germans and French built cast-lead pipe, but excessive cost prevented its general use. Hollow-bored logs also were employed (Fig. 131), but rot soon destroyed their usefulness, and it was realized than an inexpensive and durable pipe was required.

Records indicate that the earliest iron pipe consisted of converted iron cannon laid end to end and joined, and this ingenious conversion of "swords to plowshares" proved so satisfactory that city architects and planners began to suggest the casting of pipe in iron. Such pipe was supplied by the founders of Europe.

The earliest true water line made of cast iron is said to have supplied water for Dillenburg Castle in Germany, and was installed in 1455. Earlier installations may have existed, however, since there is mention of iron-pipe founding at Siegerland in that same year, and German waterworks were famous throughout Europe not long after that date. There also is a record of a line laid in 1562 at Langensalza to supply the Jacobie and Rathaus fountains. Cast-iron pipe was laid at Versailles in 1664 by order of Louis XIV of France (Fig. 132), and five miles of this line still are in use today after nearly three hundred years of service.

England, and later the United States, followed Germany, France and other European countries in the use of cast-iron water systems. In the early 1800s the demand in England for cast-iron water and gas lines was reported to be "amazing." America's first major cast-iron water line was installed in Philadelphia in 1817 with pipe imported from England. Even earlier, an 84-foot test line cast in New Jersey had been laid down, and Albany, New York, seems to have advertised in 1813 for "one hundred tons of cast-iron conduit"—but there is no record of the actual installation.

At the outset, cast pipe was molded horizontally; soon afterward, vertical pit casting came into use. The development of centrifugal or spun ferrous casting is traceable to pipe production, the first spun pipe being made in 1902 by an American named Custer at the Tacony Iron Works, Philadelphia. As a result, a new foundry procedure of great possibilities was born, and today most cast-iron pipe tonnages are produced by the spinning process.

The following quotation from the diary of an American pipe foundryman of 1819 surely is expressive of the general uncertainties that beset not only those engaged in pipe founding but also anyone involved in the casting of metals:

"Sunday . . . . Made one good pipe—lost another.
Monday . . . . Pipe good this evening.
Tuesday . . . . Made no pipe account rain.
Wednesday . . . . . . No Pipe today—bad iron.
Thursday . . . . Iron good—made two pipes.
Friday . . . . Made no pipes—iron bad.
Saturday . . . . Made two pipes."

In the fifteenth century, European iron casting was entirely dependent on direct-furnace metal—that is, castings were poured directly from the blast furnace, which also gave pig iron (Fig. 133). The word "cast," meaning to shape metal by pouring it molten into

132 *Flanged cast-iron pipe laid at Versailles, France, in 1664 and still in use today. Note initials "LF" cast on pipe, for "Louis of France" (Louis XIV, King of France, 1643-1715).*

a mold, seems to date back to the late 1100s and to have had its origin in the old Norwegian word *kasta.* The word "foundry," however—the place where the founder casts his metals—dates only from the late 1500s and apparently came into use in connection with operations of the early ironfounders who poured metal direct from the blast furnace.

Actually, the word "foundry" is derived form the French *foundre* and the Latin word *fundere,* both meaning "to pour." An interesting old form of the word appears in the following description of blast furnaces and early iron foundry operations, by John Ray of England, in 1672:

"The hearth, or bottom of the furnace, is made of sandstone, and the sides round, to a height of a yard; the rest of the furnace is lined up to the top with brick. When they begin upon a new furnace they put a fire for a day or two before they begin to blow. Then they blow gently and increase by degrees until they come to the height in ten weeks or more. Every six days they call a founday in which space they make eight tun of iron, if you divide the whole sum of iron made by the foundays; for at first they make less in a founday, or last more."

In other words, "founday" was "pour-day," which in the plant Ray described was every sixth day.

This type of blast furnace, of course, was the forerunner of the cupola, although a change in fuel had to take place before the cupola could be invented. All blast furnaces up to the early 1700s were fired with charcoal. In consequence, forests of the European continent were being stripped by the hungry furnaces at a tremendous rate. The call for guns, stoves and pipe was so great that whole areas were denuded of trees within a few years.

Lack of wood soon began to threaten, and finally to strangle, the blast furnaces, both in England and on the Continent. As a result, stringent laws were passed governing the cutting of further wood and the building of additional iron works. The wood shortage became so acute that all but a few furnaces closed down, and England was forced to import iron from Europe. However, by 1750 coke had come into general use and England, then Scotland, both rich in iron and coal, went back into iron production; by 1788, English pig iron production had reached 68,000 tons annually.

The history of the use of coke deserves full mention. We know that the early Chinese, and probably the Greeks and Romans as well, used coal; but coal as a fuel was long unknown to the Europeans and was not adapted to industry until late in medieval

133 *Scene in an early iron foundry, depicting the pouring of direct blast-furnace metal into sand molds for stove-plate work.*

times. We cannot be certain who first converted coal into coke, although a German, Simon Sturtevant, appears to have secured in 1612 an English patent on a method of smelting iron by means of "sea-coale, pit-coale, and earth-coale."

The first notable attempt to use coal in ironmaking, however, was that of Dud Dudley, an English ironfounder of 1619, who conceived the idea of firing a blast furnace with coal. Dudley lived in Worcester, England, and in 1619 obtained a patent from King James for smelting iron with pit coal. In addition, he used a larger than

134 *Small remelt furnace developed by the French about 1700, similar to the experimental furnace used by Reaumur for making whiteheart malleable iron.*

usual furnace and bellows and obtained about seven or eight tons of pig iron per week. This forward step to save England's timber supply and open a great industry was effectively blocked by a combination of charcoal ironmasters, who throttled it for a hundred years until economic circumstances forced their hand.

Fortunately, Dudley was not silent. Driven out of business time after time, he wrote a book in 1665 on the procedure, which he called *Metallum Martis*. This caused other ironmasters to give more thought to fuel; and as the fuel shortage became acute, reference to this book aided them materially. Particularly was this the case with one Abraham Darby of Coalbrookdale, England, in 1730.

Darby first mixed charges of charcoal and raw coal; then, realizing that something else was needed, he treated the coal as his charcoal burners treated wood. The result proved quickly successful, and he erected seven of the first coking furnaces at Coalbrookdale. With the use of coke he was able to make coke iron at about two-thirds the cost of charcoal iron, and thus was developed one of the principal tools of the iron foundries.

135 *Iron foundry operations some two centuries ago. Metal was channeled to reservoir and ladled out from there. Weight on tackle was shifted from mold to mold while pouring. Curtain on outside was dropped in the event of bad weather.*

With a good supply of hot fuel and a widening demand for iron castings, foundrymen began to experiment for better metal and cast metal properties. About 1700, the French founders conceived the idea of remelting pig iron in a separate, smaller furnace (Fig. 134), and this was the great step that eventually brought ironfounding into the field of metallurgical chemistry (Fig. 135).

As has frequently happened in history, a great man was available to cultivate this opportunity. He was Rene Antoine de Reaumur (1683-1757), the great French physicist and naturalist, who recorded much of his important experimental work. First came the rebirth of crucible steel in Europe. Here Reaumur carried out in crucibles a process of making steel wherein the metal was reduced from the ore and carburized in one operation.

Even more important was his method of annealing hard-iron castings. He heated castings embedded in iron ore and proved that in such heating much of the brittle character of the metal was removed, rendering it easier to tool. Although somewhat limited to small castings, Reaumur developed the first malleable iron, known today as "European whiteheart." It received its name from the gray or white color of the fracture, as contrasted with the "American blackheart" malleable iron which came a century later. The French cupola for remelting iron, and Reaumur's process of annealing, did more for iron-foundry progress in twenty-five years than any process of the previous four hundred years.

Reaumur's melting furnace was quite small and the annealed castings were at first limited in size, although new foundry progress coming rapidly in all quarters soon removed this limitation. One unexpected development of the small French remelt furnace was the establishment of a most curious trade, the traveling foundry on wheels, exemplified by the tinker molder who traveled over England, making odd castings in his small furnace which he charged with scrap or whatever was available. To this unsung itinerant foundryman credit must be given for spreading the idea of a separate furnace for the remelt of pig iron and the making of additions—which was to lead the way to the chemistry of iron.

These little traveling furnaces were about 24 inches high and 9 inches in diameter. They were made in two sections, and the entire furnace was tilted to pour.

In 1794 appeared the first cupola, similar in appearance to those of today, invented by John Wilkinson of England, who used the principle of the portable furnace but built it in blast furnace size. He placed the taphole at the floor line rather than at the tuyere level. To provide blast he used for the first time the steam engine invented by James Watt in 1765. Wilkinson also was the first to use a cast-plate construction for the furnace shell. This cupola melted cast gun-borings with charges of coke. Iron foundrymen everywhere realized from the new castings thus obtained that great progress had been made (Fig. 135).

136 *Melting steel by the crucible process, as developed in England by Huntsman, about 1750. It is known that crucible steel was cast in India around 500 A.D., but the process appears to have been "lost" for over a thousand years before Huntsman's "rediscovery."*

Although an Englishman is credited with the cupola's invention, the origin of the word "cupola" is German, with a Latin derivation. This was due to the fact that the German blast furnaces were fitted at the top with a short tub having no bottom, for ease in charging. The Latin for cask is "*cupa*," which became "*Kuppel*" in German, and finally "cupola." Therefore, the foremost piece of equipment for the founding of iron is not, as is usually the case, named for its inventor or its purpose, but rather for a charging device taken from the blast furnace.

In addition to malleable and cupola cast iron, still another ferrous metal now gained attention—a metal reintroduced after long absence from historical records. About 1750, a Benjamin Huntsman announced the "official discovery" of crucible cast steel in England (Fig. 136). With the exception of some experimental steel developed by Reaumur, all steels at that time had been of the cementation type—wrought bars decarburized by the ancient process of packing in charcoal. Huntsman's discovery, however, consisted mainly in the development of a crucible of superior refractory properties. The age-old method of treading the clay was still used (Fig. 137), but Huntsman achieved a combination of clays which could be combined so as to withstand the necessary temperatures of molten steel. His crucibles were said to have held 34 pounds.

We know, of course, that the eighteenth-century development of crucible steel was antedated by the Hindus centuries earlier, but such is progress when a world exists without adequate communication and exchange of information. Nevertheless, steel in its cast form was not to achieve real importance for another hundred years. As important as were Huntsman's discoveries, the quantity of steel produced was small, and the process was expensive.

Some mention also should be made at this point of progress in molding during this period of growth. In 1709 we again find Abraham Darby, even prior to his development of coke, working on a new device that proved to be the first true foundry flask for sand or loam molding. All molding, with few exceptions, had been carried on previously in pits, on the floor, by use of pattern boards tied together, or in crude box frames.

Unfortunately, Darby and his assistant, Thomas, guarded their work extremely well, and even today it is not known exactly how their flasks appeared. Enough data have come to light, however, to make it certain that flask equipment was initiated at this time in a crude way in the molding of cast-iron cooking pots. It is known that Darby obtained a patent of meager descriptive terms for making dry-sand molds in multi-parted boxes, but the description did not reveal exactly how it was done. The process remained a mystery for nearly a hundred years, but the few known facts on Darby's flask finally joined the cupola in supplying the ironfounder with the tools essential to building the vast mechanical world that castings were to make possible (Figs. 138, 139, 140).

137 *Treading clay for the manufacture of crucibles, an age-old method still in common use as late as the middle of the eighteenth century.*

138 *Early uses of special foundry flasks for the production of small cast objects free from cores. The first true foundry flask has been credited to Darby of England, in 1709.*

139 *Two-part flask for bell molds. Left: two box parts. Center: rammed up cope. right: complete mold in flask. (See also Fig. 140.)*

140 *Same flask as in Fig. 139. Left to right: stripping cope; pattern; drag; mold assembled for casting.*

Iron and castings, by the end of the eighteenth century, were coming to the fore in all industries. Particularly were they being widely adopted for the infant steam engine and in the budding railway industry. Castings of iron were used for the first rails in English mines in 1767. The first flanged rail was a casting, as was the first cast-iron car wheel made for mine cars, in 1800. A chilled-iron plowshare provided the impetus for another foundry process.

Blast equipment, notably advanced in 1769 by John Smeaton's compressed air pump, was further improved by the Newcomen engine and, later, by the Watt-Boulton engine. Cast iron had become the subject of more exact testing, and knowledge of the metal and its treatment had broadened materially. As a result, increasing interest was shown in iron by scientists, inventors, engineers, industrialists, artisans and builders.

From doing yeoman service as firebacks, grave markers and cannon, cast iron now became the material of the architect and, later, the machinery builder. By the 1800s it led all cast metals in tonnages produced. Widely used by now in construction, cast iron had supplanted the timbers in bridges. It shaped "rail courses for wheel carriages," wheels for mine cars, railroad rails, locomotive parts. It went into columns, bearers, cylinders, rolling-mill rolls, machine frames and myriad parts, crankshafts, ingot molds, furnaces, boilers, frames and parts for forge hammers, gears, mill wheels, lock gates, barges, ornamental iron, aqueducts, hydraulic engines, plowshares, jewelry, toys, baptismal fonts and statuary—a tremendous list of products and commodities, including greeting cards. An early eighteenth-century writer made the statement that "there is hardly a single article capable of being manufactured in wrought iron, the like of which the ingenuity of the sand molder cannot produce in cast metal."

John Wilkinson had great faith in iron and its usefulness. Before he died in 1800, he saw that great faith justified, with iron being applied wherever "a durable material was required." He was buried in an iron coffin, and an iron tombstone marked his grave.

Meanwhile, across the Atlantic, a new nation was rising at this time, founded by men who were vigorous, mechanically minded and resourceful. The western march of man having colonized the Americas, the development of castings followed closely. To the New World in the west came the foundry practice of the Old World, developed through centuries of accumulated experience in Europe and in Asia.

CHAPTER 8

Colonial Foundries in the New World

VIRTUALLY no foundry developments differing materially from those previously brought forth in Europe and Great Britain can be found in the Americas prior to 1800. However, the tremendous increase in the use of castings which developed in this country after that date makes it imperative to trace the growth of the infant foundry industry in the Western Hemisphere.

Prior to colonization of the Atlantic seaboard of the United States and Canada by the English, Dutch and French in the sixteenth and seventeenth centuries, castings had been produced by the native inhabitants, but not generally in North America. The North American Indians had no knowledge of metallurgy, according to present research, except for some facility with the cold working of copper.

There is some evidence that the prehistoric mound builders, who preceded the Indians, might have carried on some work with melted metals. The mounds located in what is now Ohio and the Mississippi Valley have yielded certain castings. Undoubtedly the same crude methods were employed as were used by the prehistoric peoples of the Middle East.

For all practical purposes, however, the New World was only beginning to emerge from the Stone Age when the white men first landed on these shores, bringing with them their knowledge of cast metals. Such was not the case, of course, with the more advanced native civilizations of Central and South America, where in Peru, Colombia and Mexico intricate examples of gold and silver casting have been discovered. It seems quite clear that the inhabitants of these areas already had developed for themselves the lost-wax method of molding, for in no other way could some of the complicated cast pieces attributed to them have been made.

Gold, however, was the primary metal; there is no evidence of the use of iron. As a result, the civilization of the South and Central American areas was restricted for, as we have seen, the foremost sign of a fully progressive people lies in the melting and casting of iron or alloyed non-ferrous metals. Without these, apparently, no civilization has ever progressed far.

Records indicate that, as a general rule, this hemisphere was explored for gold but colonized with iron. Iron first made its appearance as a result of a deliberate search by Sir Walter Raleigh, who advised of the finding of iron ore deposits on Roanoke Island in South Carolina. Samples were sent back to England, but no action was taken on their development for nearly forty years. In 1607 the first colony at Jamestown, Virginia, was established, and again iron was sent back for analysis.

It was not until 1622 that there was an actual attempt to make use of this mineral. In that year an iron blast furnace was established at Falling Creek, Virginia, near Richmond, with skilled melters and foundrymen from England—an enterprise that was completely wiped out by an Indian massacre prior to the time the furnace went into operation. Since none apparently lived to tell what happened, details of the project are missing (Fig. 141).

It fell to Massachusetts to have the honor, in 1642, of becoming the birthplace of the first American casting, the famous "Saugus pot." This original American foundry was established near Lynn, Massachusetts, on the Saugus River, and has been referred to in history as the Saugus Iron Works. Some details of this operation give us a picture of the typical iron foundry of that period.

The founders of the enterprise were well-matched partners. Thomas Dexter, the mechanic and builder, was a rash and outspoken man of direct action. In 1633 we find him sentenced to sit in stocks for "seditious talk." His partner, Robert Bridges, the exact antithesis, was a pious and exemplary man and promoter of the project.

It was Bridges who took samples of Saugus area bog ore to England and obtained the necessary financial help for starting operations. Thus was founded the company known as "The Company of Undertakers for the Iron Works," with capital of £1000, which founded the small village of Hammersmith, so called because the imported furnace and foundrymen came from the town of Hammersmith, England.

On the fourteenth of October, 1642, a grant was made to the company which is of great interest to present-day foundry operators, as it indicates the extent to which government can go in encouraging private enterprise. Saugus Iron Works was granted the exclusive right to make iron for twenty-one years, during which time it could freely mine ore, cut wood, dam streams and set up furnaces, and was given public lands on which to operate tax free. The firm was allowed to sell and transport freely, and all employees were completely exempt from military duty of any sort. Members of the company could even stay away from church without losing their voting privileges!

With such a start the company built a four-sided hollow stack 20 feet high, 24 feet square at its base, 6 feet in diameter at its top and with an interior diameter of 10 feet. The blast was operated by a water wheel which could not operate in freezing weather; thus no winter operations were possible. Bog ore dug from neighboring swamps was charged into the furnace, together with oyster shells for flux and charcoal for fuel. Capacity was estimated at eight tons of iron a week.

141 *An Indian massacre at Falling Creek Furnace, Virginia, in 1622, wiped out the first attempt to establish a blast furnace in the American Colonies.*

142 *Joseph Jenks, master molder at the Saugus Iron Works, Lynn, Massachusetts, exhibiting the first iron casting made in America—the famous "Saugus pot." The Saugus Iron Works failed in 1688 and went out of business.*

It is interesting to note that the first metal of the new plant was made into a casting, illustrated in Fig. 142. The company had retained one Joseph Jenks as master molder, who molded a cooking pot in a small mold buried in a hole in the ground, the resulting casting weighing about three pounds, with an internal diameter of about 4½ inches. Fortunately this "Saugus pot" casting has been preserved and now is the treasured property of the city of Lynn (Fig. 143).

Thus was ferrous casting on this continent commenced a little over three hundred years ago. It introduced an industry which was destined to grow in importance until, as a basic part of our economy, it now represents an investment of billions of dollars, a source of employment for millions and a supplier, either directly or indirectly, of most of our daily needs (Fig. 144).

Some information on labor and working conditions at this earliest of American iron foundries may be of interest. Labor at the plant was made up mainly of Scottish prisoners of war captured at Dunbar by Cromwell in England. Two other types of laborers also were available: first, indentured men who had sold their services for seven years for passage to America, and, second, freed men. Occasionally labor was even recruited from the Indians during rush periods.

143 *Photograph of the historic iron "Saugus pot" and memorial plaque to this "first casting made in America," now the property of the city of Lynn, Massachusetts.*

144 *A pig of iron from the Saugus Iron Works—the first pig iron produced in the western hemisphere, so far as records show today.*

Pay averaged from ten cents per hour for unskilled labor to seventeen cents per hour for skilled labor. A work day was from sunrise to sunset, six days a week, and the crew at Saugus Iron Works probably averaged about fifty men.

Saugus Iron Works should be remembered on several historic scores (Figs. 145, 146). Jenks, the first molder, obtained the first patent granted to anyone in the colonies for the invention of the two-handed scythe, a tool still made in the same original shape. Jenks also was responsible for coining the first American money—the Pine Tree coinage of the Colony of Massachusetts. Also, due to the interest of John Winthrop, Jr., son of the governor of Massachusetts, the first fire fighting equipment for Boston was made in this plant in 1654.

Although it formed the start of an industry which eventually was to number over five thousand plants, the Saugus Iron Works itself never became great. The company failed in 1688 due to litigation, nuisance suits and reduction of timber resources; yet it furnished an important beginning for the new colonies. It should be a matter of pride to foundrymen of today that the first industrial plant in America was a foundry and occupied an important place in the industrial history of the United States.

From this early beginning the direct-iron blast-furnace foundry spread quickly. A second works was started at Braintree near Boston in 1645, and we have this verbatim report on the plant as it existed in 1647: "We have cast some tuns of pots, likewise mortars, stoves, skillets, etc. Our potter is moulding more at Brayntree as yet, which place after another blowing we shall quit, not finding mine [ore] there."

145 *Method of obtaining bog from swamps, a method extensively utilized for the early iron furnaces of the American Colonies.*

In quick succession, plants were established at Taunton in 1653 and at Concord and Rowley in 1657. By 1700 a dozen plants existed in eastern Massachusetts, including that of the Leonards at Raynham, and Massachusetts became the center of metalworking. Meanwhile, activity had spread to other states. John Winthrop, Jr., built Connecticut's first blast-furnace foundry at New Haven. In 1658 Joseph Jenks, Jr., son of the famous first molder at Saugus, erected a plant at Pawtucket, Rhode Island, which burned out in 1675. Henry Leonard, who was from a famous foundry family, moved into New Jersey and established the industry in that state.

146 *The Old Ironworks House, Lynn, Massachusetts, one of the oldest in the United States, with a framework of oak imported from England. It once housed workers of the original Saugus Iron Works.*

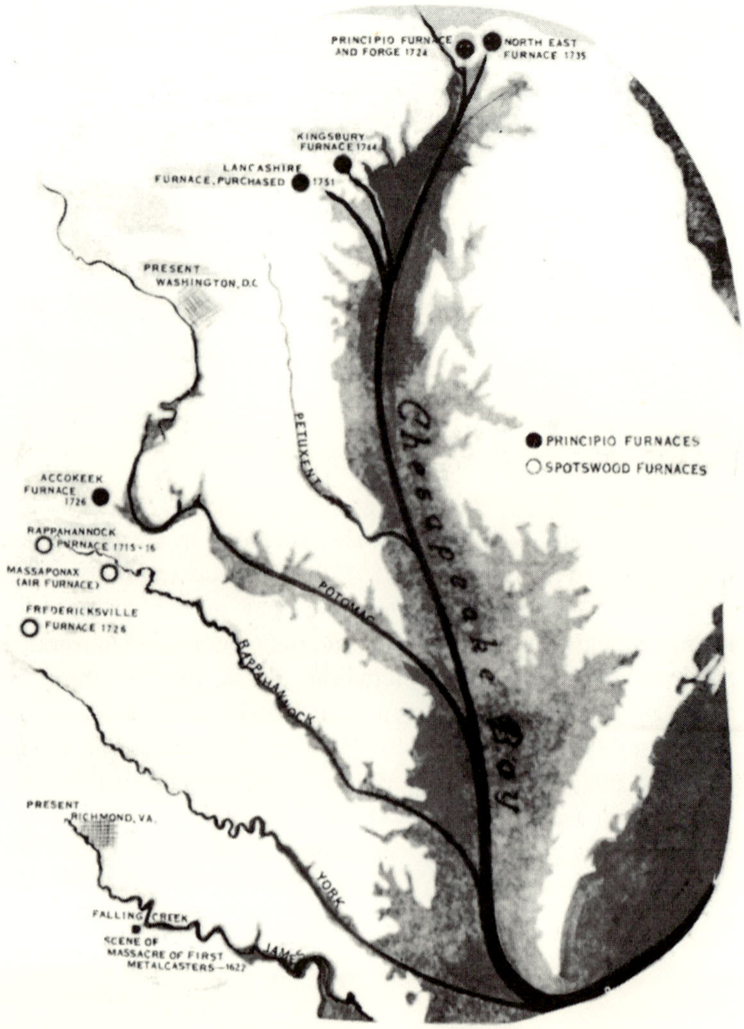

147 *Growth of iron furnaces in the Delaware, Maryland, Virginia area. Note the Principio Furnace and Forge (top of map), whose founders included Augustine Washington, father of George Washington.*

Col. Lewis Morris was also an early operator there and is said to have employed the first cast-iron-cylinder compressed air blast in America.

In the southern part of Maryland and Virginia no iron was produced for many years following the abortive attempt at Falling Creek. However, Virginia in 1715 was blessed with a progressive governor, Col. Alexander Spotswood, who labored steadily to forward iron foundry progress (Fig. 147). He established a furnace on the Rappahannock River at the junction of the Rapidan. Pig iron was hauled fifteen miles to Massaponax where, by means of an air furnace, the metal was cast into firebacks, cooking utensils, andirons, etc.

148 *Early Colonial forge operated by water tower of Italian design. Forge hammer is operated by waterwheel.*

149 *Diorama of cold-blast furnace making pig and cast iron in Colonial America. Note water-drive bellows and charging bridge.*

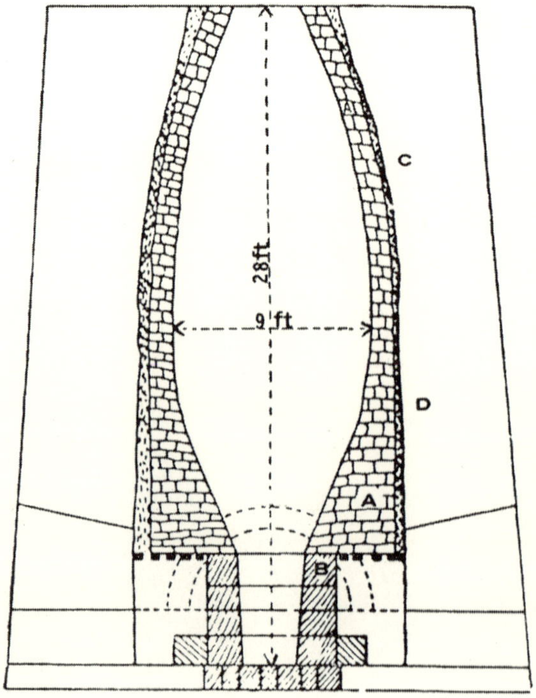

150 *Cross-section of blast-furnace foundry installation at Lakeville, Connecticut, in 1762. (A) Brick lining of shell. (B) Crucible. (C) Outer lining of shell. (D) Furnace exterior. (See also Fig. 151.)*

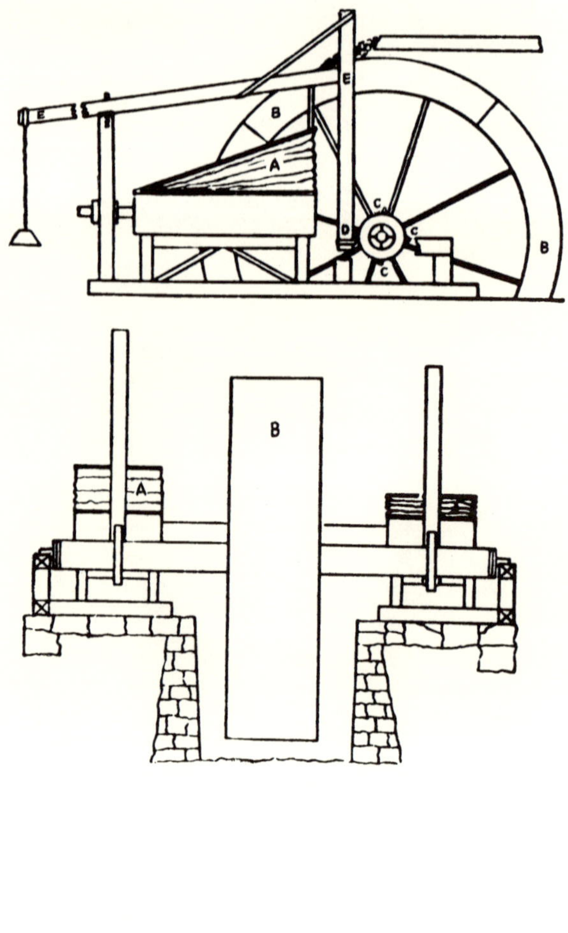

151 *Blast furnace blowing apparatus in section. (A) Bellows. (B) Waterwheel. (C) Struts. (D) Counterweight trigger. (E) Counterweight level. (Same installation as Fig. 150.)*

152 *Charging a blast furnace by hand from the charging bridge. While this method of charging closely resembles early Colonial operations, the furnace is of a considerably later type.*

153 *Early American fireback cast at Lynn, Massachusetts, about 1660.*

154 *American fireback castings of the seventeenth century. This casting made from a full pattern. On page 173, the only decorations show initials of maker and date.*

155 *Colonial stove made by Baron von Stiegel at the Elizabeth Furnace in Lancaster County, Pennsylvania, about 1759. Fuel door was located at the back.*

The quality of the iron produced by this remelt process surprised the English and is truly remarkable when it is recalled that the remelt process had only just begun in France. Spotswood established another furnace at Fredericksville, Virginia, in 1726, but his efforts never reached the limits of his planning. Figs. 148-155, inclusive, illustrate several early American furnaces and products.

To the north, the father of George Washington was engaged in establishing the Principio Company, which was to exist for two hundred years. Augustine Washington built a furnace in 1724 at the

156 *Cast-iron stove presented to the State of Virginia in 1770 and now preserved at the state capitol in Richmond.*

157 *Specimen of pig iron produced at the Union Furnace at High Bridge, New Jersey, in 1742.*

head of Chesapeake Bay in Maryland, and his company built a chain of plants in rapid succession in and around Baltimore and in northern Virginia. By 1750 Maryland had eight operating furnaces.

Such furnace expansion continued until 1754 and was successful for years as a commercial foundry and furnace enterprise. New Jersey and Pennsylvania had by then entered the race to produce iron and other castings so badly needed in this new and rough country, and were prominent iron producers by the middle eighteenth century (Fig. 156).

One early New Jersey plant is particularly interesting in that it still exists as a foundry and traces its lineage back to 1742. The Taylor-Wharton Iron & Steel Company of High Bridge, New Jersey, is without question the oldest foundry and probably the oldest industrial corporation of continuous existence in the United States.

This plant had its beginnings in December of 1742 when a foundry furnace was erected by William Allen and Joseph Turner under the name of the Union Furnace (Figs. 157,158). Shortly afterward, in 1754, another furnace was built in the same area, called the Amesbury Furnace (Fig. 159). In 1760 Robert Taylor became works manager of Union Furnace and subsequently took over the plant in 1780. During the Revolution this furnace supplied the Colonial troops with guns and shot. In 1860, still by direct descent, the name was changed to Taylor and Lange. In 1868 it became Taylor Iron Works, and finally adopted the present name of Taylor-Wharton Iron & Steel Company in relatively recent years.

Pennsylvania also was moving rapidly to exploit her mineral resources. Again iron took the lead, due to the needs of the expanding frontier and the fact that the inhabitants relied on iron in carving their homes out of the wilderness.

Several interesting iron works in eastern Pennsylvania deserve attention. In 1742 the Franklin stove was invented by Benjamin Franklin, who obtained his castings from a foundry known as the Warwick Furnace, near Warwick, Pennsylvania. This stove, a new invention soon widely adopted, was made possible through castings and materially increased cast metals tonnage.

In 1742, for example, a furnace-foundry known as the Mount Joy Forge was erected in Chester County, Pennsylvania, on Earl Valley Creek; later the name was changed to Valley Forge. This famous foundry was burned by the British in 1777 and, as every American schoolboy knows, Valley Forge served as the encampment of the American Army during that terrible winter of 1777-78 when the American Revolutionary fortunes were at their lowest ebb. Later this furnace was rebuilt, and important early experiments on steel castings were carried on there. American history, indeed, is closely and frequently associated with the foundry industry.

158 *Ruins of Union Furnace at High Bridge, New Jersey, the original plant of the present Taylor-Wharton Iron & Steel Co., probably the oldest industrial corporation of direct lineage in the United States.*

159 *Ruins of Amesbury Furnace in New Jersey, built in 1754 close to the Union Furnace. The Amesbury Furnace became the Taylor Iron Works in 1808 and is directly traceable to the Taylor-Wharton firm of today. (See text, page 175.)*

160 *An eighteenth-century "iron plantation" at Hopewell Village near Birdsboro, Pennsylvania. Such plantations, great estates devoted exclusively to iron production, existed primarily in eastern Pennsylvania in the eighteenth century. The Hopewell project now is being restored as a national park.*

No history of the American foundry industry would be complete without a description of the iron plantations, great estates which existed principally in eastern Pennsylvania in the eighteenth century. Dozens of these semi-industrial and partly feudal establishments, "growing" iron rather than cotton, had been established by 1750. One typical enterprise was the Hopewell Village near Birdsboro, Pennsylvania (Figs. 160, 161), now being restored by the National Park Service and well worth a visit by any foundryman.

William Bird bought this tract of possibly 10,000 acres in 1743; in 1761 his son built the furnace and developed the plantation. At one time nearly 1,000 persons (including furnace men, molders, miners, charcoal burners, wagon-men and their families) lived there and derived their existence almost entirely from the production of pig iron and castings. The iron mine was located a mile or so from the furnace, and ore was carted in. The furnace used charcoal in stupendous quantities, averaging around 15,000 cords a year (Fig. 162). Most of the inhabitants were wood cutters and charcoal burners, who prepared the charcoal in mounds in the forest.

161 *Key to Fig. 160, the Hopewell "iron plantation." (1) Mansion house. (2) Bake oven. (3) Spring house. (4) Barn. (5) Carriage house. (6) Corn crib. (7) Office. (8) Charcoal storehouse. (9) Furnace bridge. (10) Mill wheel. (11) Furnace. (12) Casting house. (13) Ore roaster. (14) Wheelwright shop. (15) Blacksmith shop. (16) Slag. (17) Dam. (18) Schoolhouse. (19-22) Tenant houses. (23) Tenant barn. (24) West head race. (25) East head race. (26) Tail race outlet.*

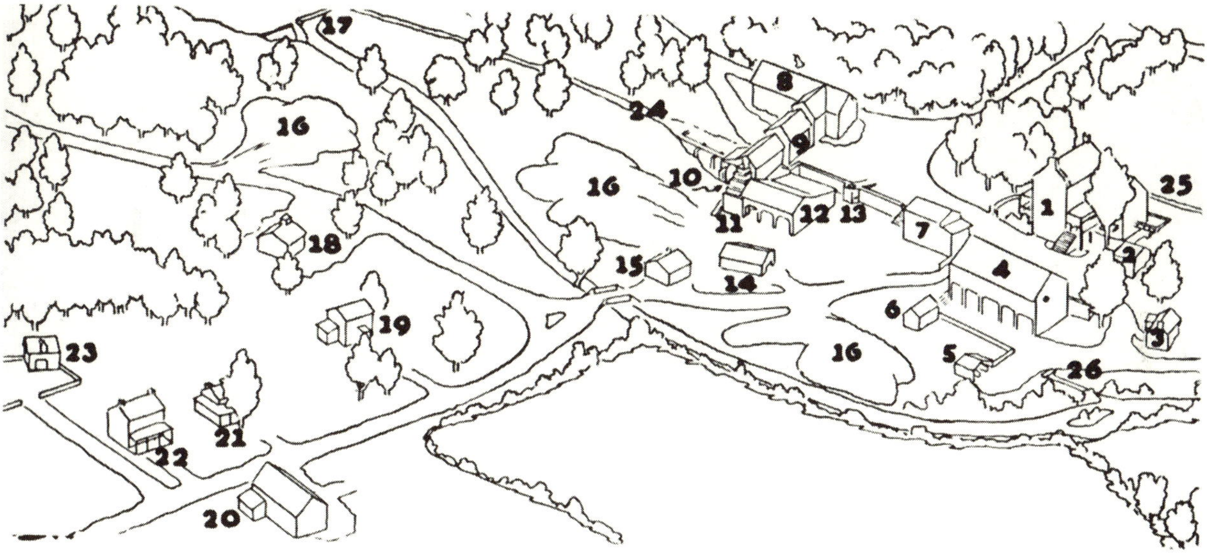

Iron and castings were sent to Philadelphia by boat or cart, but the inhabitants lived, worked and died on the plantation. The workers were free, of course, but for the most part were content to live out their lives on the plantation. The patriotism of the younger Bird to the American Revolutionary cause cost him his plantation when, in 1777, he equipped and led an entire company of 300 men to fight for the colonies and in addition loaned the Revolutionary government large sums of money, the non-repayment of which finally ruined him.

Many colorful figures developed on these iron plantations. One was Baron Henry William von Stiegel, who in 1752 operated Elizabeth Furnace in Lancaster County, Pennsylvania. (Note the stove by Stiegel shown in Fig. 155.) He produced a large quantity of castings, from kitchen utensils, firebacks and stoves to complicated pieces for export to sugar refineries in the West Indies, and became so wealthy that in 1762 he built an entire town named Manheim, where he erected a glass factory.

Stiegel brought expert German glass designers from Germany at great expense. Here was developed the famous and now prized Stiegel glass, made possible by a successful foundry enterprise. However, over-expansion ruined Stiegel, like so many others, and, in greatly reduced circumstances, he was finally forced to work as a clerk in one of his own furnaces, then in the hands of his creditors.

Toward the end of the eighteenth century, the furnaces and foundries of America began to move westward. The first ferrous foundry established west of the Alleghenies was built in Fayette County, Pennsylvania, in 1792 by William Turnbull. This foundry supplied guns and shot for General Wayne's expedition against the Indians. In the same year, the first plant in Pittsburgh was erected by George Anschultz, who made stove and grate castings. A furnace also was established at the same time on Licking River in Bath County, Kentucky. So close to the frontier was this plant that a stockade was built around the furnace, and an armed guard was required for protection against Indian attacks.

162 *Making charcoal in piles for use on a Colonial iron plantation. The Hopewell plantation is said to have consumed an average of 15,000 cords of wood a year.*

163 *Fireback of American Revolutionary War days, depicting the use of Hessian prisoners of war as common labor in the foundries of the Colonial army.*

164 *Men prominently identified with Colonial iron founding who signed the Declaration of Independence were (left to right) George Taylor, Philip Livingstone, Stephen Hopkins, George Ross, Charles Carroll and James Smith.*

Far to the north in Canada, the first Canadian foundry had been installed years before, in 1730, at a location south of Three Rivers, Quebec. This foundry operated for 150 years. As an interesting sidelight, the molders had a quaint custom of dusting visitors' shoes free of foundry dust as they left the plant. If the resulting tip was large enough, no more work was done that day, while molders, melters and the management all drank to one another's health.

A final commentary on eighteenth-century foundry operations in the United States concerns the connection between the foundry and the American Revolution (Fig. 163). It is generally accepted by every schoolboy that the stamp tax on tea and "taxation without representation" were primary causes of the American Revolution; actually, however, history reveals even more fundamental reasons, involving the casting of metals. In 1750 the English parliament, jealous of the growth of ironworking in the Colonies, passed an act prohibiting the refining of pig iron or the casting of iron in the colonies. In addition, this act restricted the building of any additional furnaces or forges. Pig iron could be made only if it was shipped to England, where a shortage of charcoal had seriously curtailed iron production.

This act, of course, would have been a fatal blow to American frontier expansion, since it meant virtual strangulation of a country badly in need of metal for its own use. As much as any other single grievance, and long before the issue of the tea tax arose, England's restrictions on the refining and casting of iron brought forth the kind of resistance that usually precedes every uprising. The act was openly flouted, and the agents of the British Crown attempting enforcement soon lost their prestige.

For the most part, the Colonial founders joined the Revolutionary cause and supported it with money, guns and shot. Other foundrymen supported it politically, and seven men prominent in Colonial iron founding (Fig. 164) signed the Declaration of Independence in 1776:

Philip Livingstone, Ancram Creek Furnace, New York.
George Taylor, Durham Furnace, Pennsylvania.
James Smith, Codorus Forge, Pennsylvania.
George Ross, Mary Ann Furnace, Pennsylvania.
Stephen Hopkins, Hope Furnace, Rhode Island.
Charles Carroll, Carrollton, Maryland.
James Wilson, Hopewell Furnace, Pennsylvania.

165 *Destruction of Valley Forge foundry in 1777 by the British. This famous forge, where Washington's army spent the terrible winter of 1777-78 and which is considered, in consequence, a shrine of democracy, was later rebuilt and again operated as an iron furnace.*

Among the many foundrymen who fought in the Continental Army as officers were Nathaniel Green, Rhode Island Furnace, who commanded at the battle of Long Island; Ethan Allen, Connecticut Furnace, who commanded the "Green Mountain Boys" and forced the surrender of Fort Ticonderoga; and Lord Sterling, Sterling Iron Works, who served on General Washington's staff.

Production of war material, of course, was the principal task of the American foundries during the War for Independence. Many foundries made shot, shell and cannon in great quantities, and it was because of the efforts of these founders that supplies kept coming to Washington's ragged and exhausted troops. Frequently, these same founders remained unpaid. As always in time of war, foundries were military objectives, and the British pointed their raids toward their destruction (Fig. 165). At times undelivered cannon and shot were buried to keep them from falling into enemy hands in case the furnaces were captured.

Foundries frequently were used as direct battlefield arsenals, as was the Warwick Furnace in Chester County, Pennsylvania. At this plant, after the battle of Brandywine in 1777, it was an unknown foundryman who confiscated all the lead clock weights in the town for bullets. The foundry eventually benefited, since it was kept busy during the post-war period making brass and iron substitute weights to replace those melted down for the rifles of the Colonial Army.

During the Revolution another stalwart figure appeared whom the average person does not ordinarily associate with the casting of metals—Paul Revere. Made legendary by Longfellow's poem "The Midnight Ride of Paul Revere," that Revolutionary patriot who rode from Boston to Lexington on the night of April 18, 1775, to warn of the approach of the British, was a foundryman by trade.

Prior to the Revolution, Revere had acquired experience in casting bronze and silver for bells and tableware, and many iron articles (Fig. 166). However, his major metallurgical experience was obtained when he was assigned by the Continental government to work with Louis de Maresquelle (Lewis Ansort), a French founder of exceptional ability. Ansort was able to soften iron by mixing his metals, and he also introduced the completely bored, solid cast gun.

Under this master founder, Revere learned metallurgical techniques which later were to stand him in good stead in his further work on the malleability of copper. After the war, Revere returned to his bell-and-fittings foundry in Boston. His diary shows typical foundry troubles, since he was constantly in difficulty with his neighbors due to smoke from his stacks.

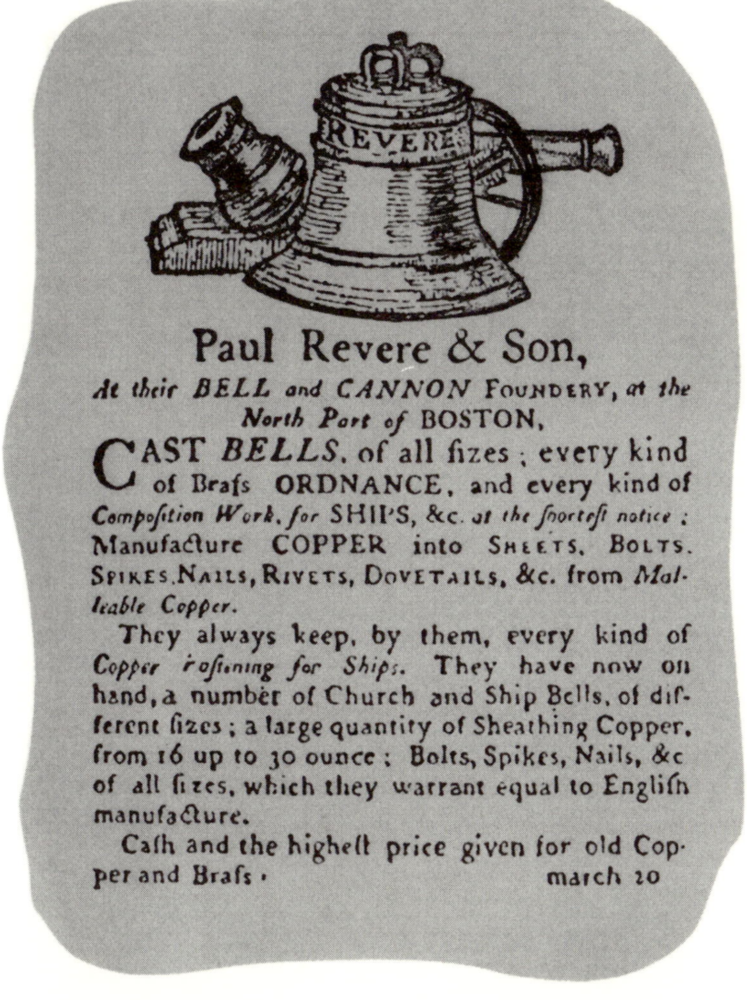

166 *Advertisement of Paul Revere's bell-and-cannon foundry in Boston. The crown on the bell dates the clipping as prior to the Declaration of Independence.*

BELL CAST IN

One of Revere's most famous bells was a recast of a bell which hung in the Old North Church in Boston and from which he had received his fateful signal "on the eighteenth of April in Seventy-five" (Fig. 167). This bell also is noteworthy for a totally different reason—from the point of view of tone, it was a poor casting. In an effort to improve tonal quality, Revere began to experiment with various types of copper and copper alloys. His metallurgical success is well known today, and a great American company bearing his name—The Revere Copper & Brass Company—is the direct descendant of Revere's original enterprise.

No account of the relation between bells and the American colonies would be complete without mention of the famous Liberty Bell, whose ringing in Philadelphia on July 4, 1776, announced the signing of the American Declaration of Independence. The Liberty Bell was cast by Thomas Lister of Whitechapel, London, to mark the fiftieth anniversary of the Commonwealth of Pennsylvania. The bell cracked twice during testing and has been recast twice, its original tone being considerably altered by the amount of copper added in the recasting. This most revered of all American bells weighs 2080 pounds and now is preserved in Independence Hall, Philadelphia, where it was originally hung.

Many Colonial founders, like Revere, cast both ferrous and non-ferrous metals. However, the Colonies also had their non-ferrous specialists. Alexander Hamilton, as Secretary of the Treasury, reporting on American manufactures in 1791, commented that "Coppersmiths and brass founders . . . are numerous in the United States." On the production of iron, he added that "The manufacturers of this article are entitled to preeminent rank."

The War of 1812 also contributed to foundry history in a rather interesting way. During that conflict Henry Foxall, a combination preacher and foundryman, was making castings for the United States at Georgetown, Maryland. The British, after burning the

167 *Bell cast by Paul Revere from metal of the bell that hung in Old North Church, Boston, "on the eighteenth of April in Seventy-five." This bell is now preserved in the nave of St. James Church, Cambridge, Massachusetts.*

White House in Washington, marched toward Georgetown to destroy the foundry, and Foxall vowed that if his foundry were spared he would establish a new Methodist church.

Due to a sudden terrific electrical storm, the British were delayed and then prevented form reaching the foundry. As good as his word, in 1815 Foxall built in Washington, D.C., the Foundry Methodist Church, today one of the largest and most beautiful churches in the capital. Many presidents and distinguished visitors have worshipped here, and the story of this house of worship calls to mind the fact that John Wesley began Methodism in London, England, many years before, in what previously had been a foundry.

The account of the development of cast metals in colonial United States ends with the infant but vigorous foundry industry spreading the length and breadth of the new land. American genius for mechanical achievement was prepared to advance the general world-wide progress of the industry. The nineteenth century, in turn, brought forth progressive mechanical and metallurgical advancements that were to accelerate greatly the slow and laborious gains already made in the art of casting metals and metal products to serve mankind.

CHAPTER 9

Metals and Melting in the Nineteenth Century

AT THE TURN of the nineteenth century, both in Europe and in the Americas, foundry practice and foundry methods had vastly improved, yet the nineteenth century was to bring forth such improvements in metals, equipment and processes as even the most farsighted and visionary foundryman of 1800 could never have imagined. The full possibilities of iron became far better understood during this vigorous century, and gray iron, the most versatile and diverse of all cast metals, was brought closer to its rightful level of expansion. As a result the use of iron for castings was vastly increased, even though malleable iron, chilled iron, and finally cast steel also were tremendously advanced during this period.

The world was approaching the highly mechanized state in which we now find it, and the iron family and its older non-ferrous relatives, readily castable, economical and serviceable, were destined to play vital roles in that mechanization. None of our modern luxuries and few of our modern essentials would be available today were it not for the foundry industry and the cast metals which the artisans, engineers and inventors of the nineteenth century made usable and practicable.

First and foremost was the true birth of gray iron, a metal little understood and still greatly underestimated. Even today, gray iron frequently is placed in somewhat the same category as the early castings of direct blast-furnace iron, which of course were extremely limited in their application. Modern gray iron is a product of the cupola and close chemical and metallurgical control, both nineteenth-century developments.

168 *Open-front, stationary-bottom cupola of the 1820s. Note stairway leading to the charging platform, up which the charge had to be carried in order to charge from the top.*

Gray iron is a true illustration of the chemistry of metal as applied to the science of casting, and the developments of the four years between 1810 and 1815 in the field of cast iron reveal this relationship. In 1810 Louis J. Broust, a great French chemist, after extensive investigations described cast iron as a solution of carbide of iron in iron. To modern metallurgists this must seem extremely elemental, but when it is realized that little scientific data was actually available then, the work of those early chemists assumes its proper proportions.

Also in 1810, Jons Jakob Berzelius, a Swedish chemist, produced ferrosilicon by melting silica, carbon and iron fillings in a sealed crucible. In that same year Wilhelm Stromeyer, a German physicist, produced several grades of ferrosilicon in more exact experiments and proved that it was silicon and not silica in the metal.

In 1814 Karl Karsten, German scientist and metallurgist, published the results of experiments which proved that oxygen does not exist as an essential ingredient of cast iron, but that the different types of cast iron are due to the different forms of the carbon content. He then described two compounds of iron and carbon, one rich in carbon and poor in iron (graphite), the other poor in carbon and rich in iron (white iron). Karsten was also one of the first to observe sulphur. His experiments showed that 0.05 to 0.25 percent sulphur in iron made the metal red short, while as little as 0.05 percent phosphorus made iron cold short.

Chemists, however clever, were limited in research, and although they opened new vistas for foundrymen and metallurgists, in final analysis it was the practical foundrymen who put to use this new-found information which existed solely in the form of pure research without practical application. Perhaps the greatest single step in this direction was the development of the cupola.

While the Wilkinson cupola, originated in England in 1794, was a great step forward, mechanically it still fell far short of the efficient and economical melting units available to iron founders today. Records show the early cupola had a stationary bottom with a front draw built on a stone foundation. The charge was carried up a flight of steps to the top, or was thrown and shoveled from one landing or platform to another until it could be charged from the top (Fig. 168).

The shell of the cupola generally was made from castings with an opening in front of sufficient size to permit raking out the slag by means of a hook. This cupola melted very slowly, and iron dripped out continuously into a reservoir in front, from which ladles were filled. To an experienced foundry eye this indicates what troubles the early melters encountered; yet they made castings, and good castings.

In the United States, the cupola was introduced around 1815. At Baltimore two of these early cupolas (Fig. 169) still were in operation in 1902, but all have now been replaced by more modern units.

This melting unit, existing on scrap and pig iron, had the effect of so widening the breach between smelting and melting that the foundry and reduction blast furnaces soon became completely separated. Merchant pig-iron producers, relieved of the duties of casting metals, went on to achieve the highly specialized skill that

169 *Early one-piece, boiler-plate cupola with stationary bottom and charging platforms built in levels. The charge was shoveled upward from platform to platform until charging level was reached. The cupola was introduced in the United States in 1815, and two such cupolas as shown here were known to be still in operation in 1902.*

today is theirs. Foundrymen, on the other hand, provided with a ready and reliable supply of scrap or pig-iron, were able to control with greater certainty many of the variables which hitherto had proved uncontrollable.

In 1850 a further important improvement was made in the cupola—the drop bottom, without which no efficient cupola could operate today. Who the inventor was and exactly where in the United States his invention first was perfected, is not known, except that the drop bottom originated in the New England area.

This innovation, so familiar to modern iron founders, seems to have marked the beginning of modern cupola design, because developments came rapidly thereafter. The stack was made smaller than the crucible and built higher to aid draft. The blast was introduced from two tuyeres on opposite sides of he cupola, and a melt could now be poured in about ten hours.

Some ten years after introduction of the drop bottom, the one-piece cupola appeared, constructed of boiler-plate casing in both crucible and stack. Then came the introduction of air by means of air chambers and blast tubes, and next the cupola designers eliminated the taper so as to provide the same diameter from top to bottom (Fig. 170).

About this time the manufacture of cupolas in the United States was undertaken by companies dedicated to further improvement. At first, this important industrial development was the work of successful melters who built cupolas for their foundry friends and neighbors. The first such patented device was the Mackenzie cupola, known as the "two-hour" cupola—the name referring to its melting time. Following this, in quick succession, appeared the Truesdale, the Laurence, and the Daugherty cupolas.

The first commercial cupola in the United States, prefabricated insofar as possible and built primarily to market, was the Colliau cupola (Fig. 171), first introduced in 1874. It was highly successful, being a fast and economical melting unit. When first introduced it presented a number of new features, among which were a hot blast, double row of tuyeres, and other improvements.

After the Colliau, another cupola was developed and still is sold by the concern which bears the name of its inventor—the Whiting cupola. John H. Whiting in the late eighties observed the need for a commercial cupola, the result of extensive practical foundry experience. He developed a cupola which employed two rows of tuyeres, the lower arranged to form an annular air inlet that distributed the blast around the entire circumference of the furnace.

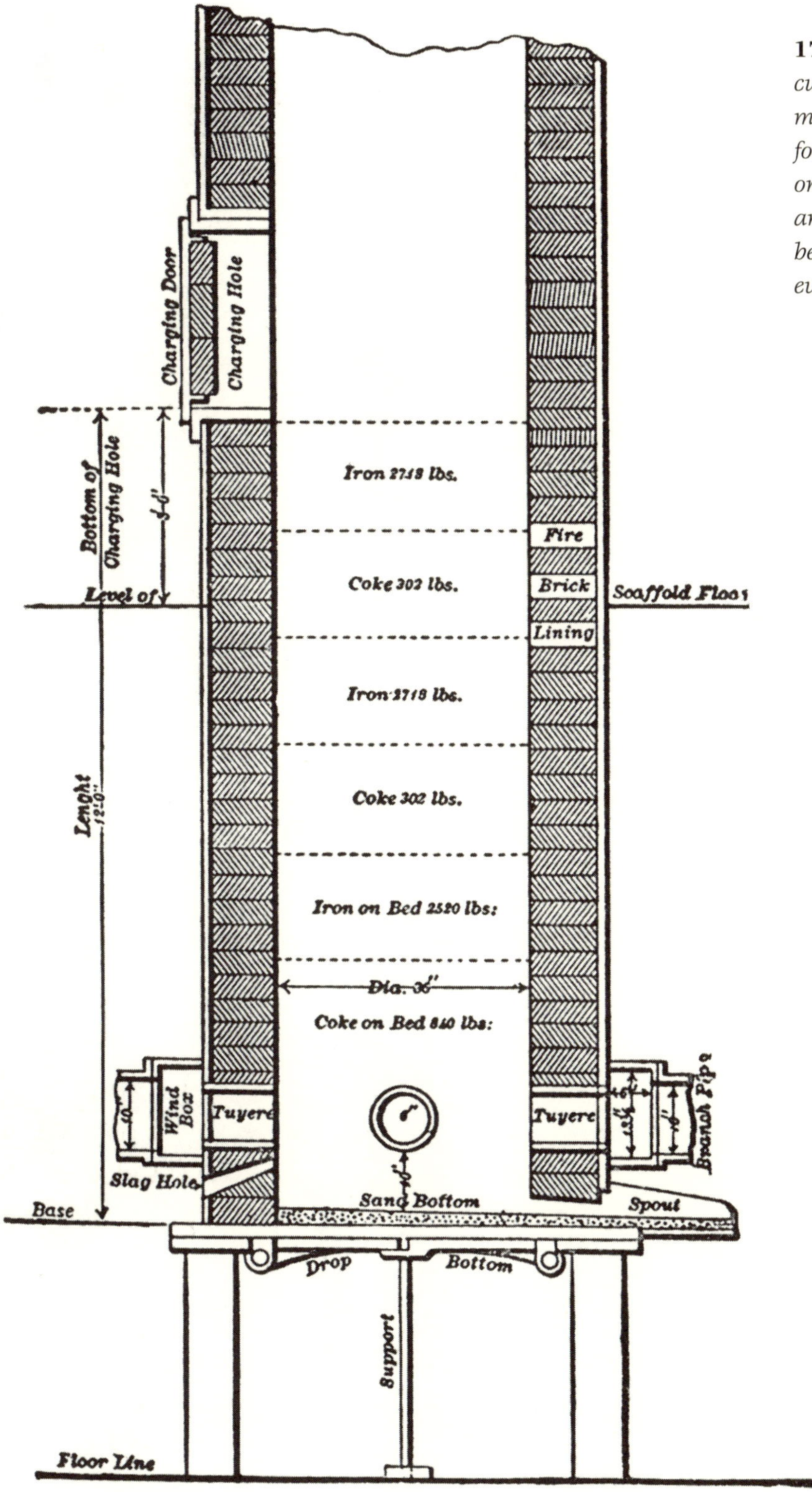

170 *The straight-side, lined cupola was introduced after the middle of the nineteenth century, following development of the one-piece cupola and air chambers and blast tubes. Cupolas have been built with straight sides ever since.*

171 *The Colliau cupola, introduced in 1874, the first prefabricated commercial cupola built in the United States. A hot blast and double row of tuyeres were among the improvements embodied in the Colliau.*

His tuyeres could also be adjusted vertically for changes in classes of work, type of fuel and diameter of cupola. The cupola permitted the use of either coal or coke as fuel and was equipped with a safety alarm and blast meter. The Whiting cupola (Fig. 172), with its standardized construction, soon proved a boon to foundrymen and was widely accepted.

The story of the cupola and cast iron in this period would not be complete without reference to the development of Connellsville coke and chilled iron—both products of the United States. Coke was in general use in Europe in 1750, but due to the heavy timber resources of the United States it was not produced here until 1817, in Plumsack, Pennsylvania. From that time until 1860, American foundrymen generally made their own coke for their own use, but thereafter coke became a commercial product, as with pig iron.

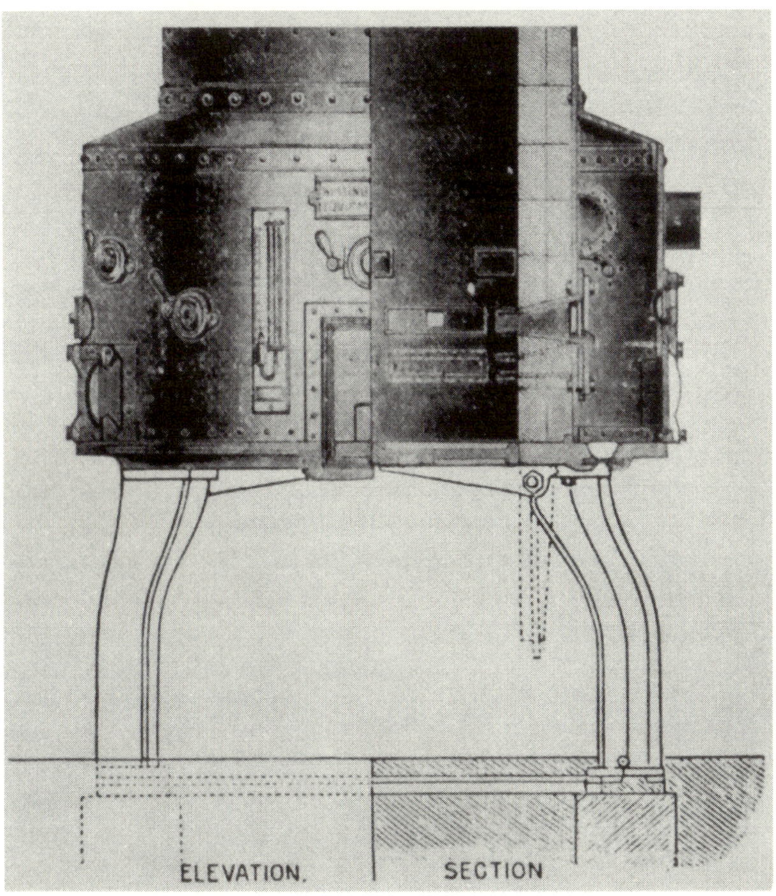

172 *The Whiting cupola, developed in the late eighties by John H. Whiting, then of Detroit, Michigan. This commercial cupola, with its standardized construction, incorporated many improvements that proved of great assistance to iron foundrymen.*

Connellsville coke was first produced in 1841 at Connellsville, Pennsylvania, in beehive ovens. The product proved so popular that the demand for Connellsville coke became and remained heavy until 1914, when by-product coke came into greater demand.

The use of coke and bituminous coal was made possible generally by the introduction of the hot-blast furnace in 1828, by James B. Neilson of Scotland. At first the blast pipes went through the furnace, later they were run through a charcoal oven, and finally they were placed on top of the furnace itself. This device increased melting output from 10 to 40 percent, so that by 1869 the production of coke and bituminous pig iron exceeded that of charcoal pig iron, and by 1875 the use of coke pig iron alone surpassed that of anthracite pig.

Another American development of the middle nineteenth century involved the introduction of chilled-iron car wheels. Asa Whitney of Philadelphia obtained a basic patent in 1847 on a process for annealing chilled-iron car wheels cast with chilled tread and flange. They were satisfactory so long as cold-blast charcoal iron was used, but trouble developed when melting practice changed from the use of anthracite coal to the hot blast.

In 1880 an attempt was made to introduce manganese for production of chilled iron, but the resulting product did not possess adequate wear properties. Finally, a small amount of ferromanganese was introduced directly into the ladle, with good results. Thus originated one of the basic products of the foundry industry—chilled car wheel iron. The excellent performance of this foundry product and its long-wearing and economical properties made possible the long-haul, heavier railroad freight loads of today.

No description of iron would be complete without giving full attention to American "blackheart" malleable iron, as contrasted with European "whiteheart" malleable iron. Credit for the progress and growth of the American malleable industry belongs to many men, but its origin is clearly indicated and Seth Boyden of Newark, New Jersey, is entitled to full credit for the development (Fig. 173).

Boyden's experiments, beginning in 1820, were based primarily on an attempt to duplicate the "whiteheart" European product. He was endeavoring to lower the cost of harness hardware and wanted a strong iron which could be machined easily. Because of a larger percentage of silicon than was available in the Reaumur process, Boyden was able to produce the strength, but not the white color, of the European product. However, Boyden (either by design or by accident) had shortened the time of anneal to six or ten days. This left free temper carbon graphitization, as opposed to Reaumur's decarburization.

173 *Statue of Seth Boyden in Newark, New Jersey, where in 1826 he first produced "blackheart" malleable iron, from which has grown the great American malleable iron industry.*

Boyden first used a crucible and then an air furnace with a capacity of about one thousand pounds. His annealing furnace was a beehive unit into which pots loaded with castings were lowered through the top. The beehive was eventually replaced by a continuous annealing furnace having a sloping floor, known as a "shoving" furnace because the pots and castings were shoved in one end and out the other.

Boyden's work was highly important, but the most notable advances in the malleable field were made by the men who followed him. Great among American malleable founders are the names of the late Enrique Touceda of the Malleable Founders Society and father of "Certified Malleable Iron"; J. H. Wittemore of Eastern Malleable Iron; Duncan Forbes, Sr., of Gunite Foundries; A. E. Hammer of Malleable Iron Fittings Co.; A. A. Pape and Harry A. Schwartz of National Malleable & Steel Castings Co.; and W. R. Bean of Eastern Malleable Iron and the Whiting Corporation. These men and their associates, aware of the valuable engineering properties of malleable iron, developed the malleable industry to its present level of prominence in the foundry picture.

Third in chronological sequence of casting developments came steel. For centuries, man had been near achievement of the manufacture of steel castings in large quantities, but inadequate equipment held him back. As we have already seen, the birthplace of the steel casting appears to have been India. Steel may have been poured in England as early as 1609, but the first certain record of steelmaking was the work of Huntsman in 1750, whose developments in crucible refractories permitted him to cast steel.

Steel produced by this process remained expensive, however, and it was not until the introduction of the converter, the open hearth furnace and, finally, the electric furnace that steel was produced commercially in quantities that rendered it economically practicable. The metallurgists, as always, were paving the way, and in 1781 Trohern Bergman of Sweden was the first to define the difference between wrought iron, steel and direct iron in terms of percentages of carbon.

At this time two points of view existed on the nature of steel. One maintained that steel contained more sulphur than iron; while the other opinion held that steel contained more iron than sulphur or other combustible matter. These two opposed theories were due to the fact that two methods of manufacturing steel were in use—one by the cementation process, the other direct from the ore.

In 1798 developments were further advanced when David Mushet of Scotland discovered that cast steel could be made by adding controlled proportions of carbon to bar iron in the crucible. This eliminated the long cementation process, since carbon now could be introduced during the melting with little additional expense. In 1839 William Vickers of England was able to make cast steel from wrought iron scrap by combining manganese oxide and carbon. These improvements slowly increased the output of steel castings.

Cast-steel guns were made at the Krupp Works in Germany in 1847, and church bells of cast steel also were made in that country about 1851. In 1851 an Austrian named Bachum developed means of casting steel to pattern and, also in 1851, another Austrian, Meyer, applied tungsten on a large scale to the production of cast steel.

Meanwhile, in the United States, cast steel was produced by the crucible process in 1818 at the historic Valley Forge foundry, but difficulties due to a lack of adequate materials—principally refractories for crucibles—caused the experiment to be abandoned. In 1831-32 excellent clay from Cumberland, West Virginia, enabled

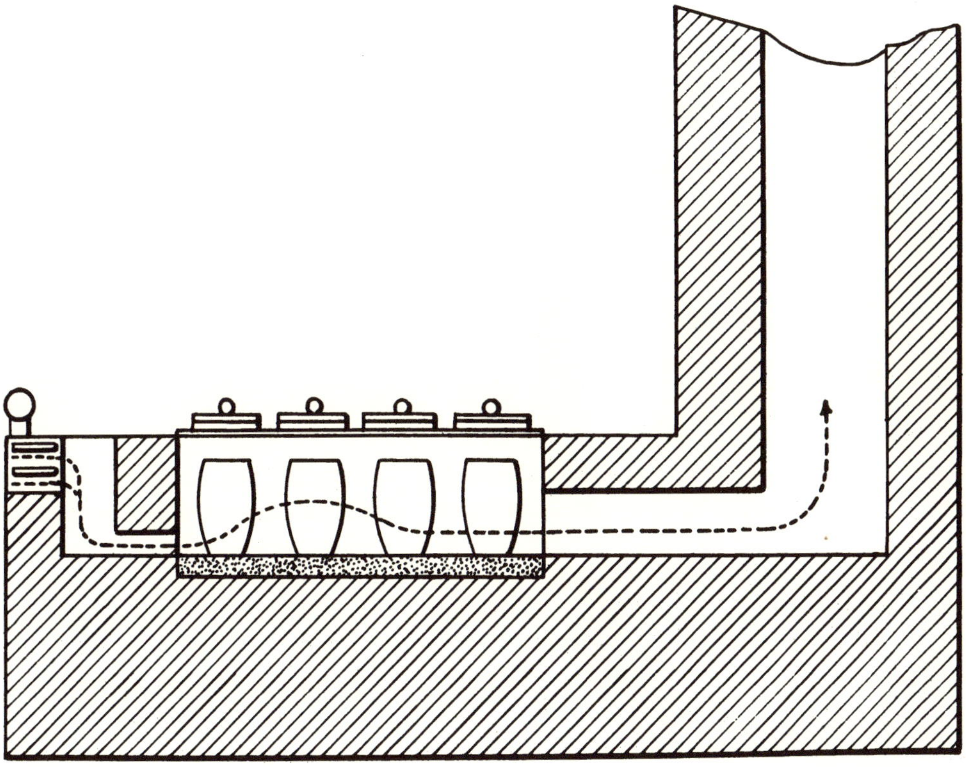

174 *Cross-section of a natural draft, oil-fired crucible furnace for melting cast steel. William Garrard of Cincinnati, Ohio, is credited with establishment of the first commercial crucible steel plant in the United States, although crucible steel was produced at historic Valley Forge in 1818.*

William Garrard of Cincinnati, Ohio, to establish the first commercial crucible steel operation in this country (Fig. 174).

It is interesting to note that the first commercial steel castings from the Garrard plant sold for 18 to 25 cents per pound and were used as blades and guards for the McCormick reaper. However, at this point a little-known American was starting experiments which were to result in the first highly successful, quantity-production steel melting unit, the converter.

William Kelly, whose invention of the converter clearly anticipated the work of Sir Henry Bessemer, although the latter gave his name to the unit, first operated his converter in 1851, a full five years before Bessemer's patent was obtained. Unquestionably, neither man realized that his efforts were being paralleled by the other, but it has been well established that Kelly was the first to use the converter.

Kelly developed his idea out of experiments on the refining of pig iron, as the result of his inability to obtain high-quality ore at his

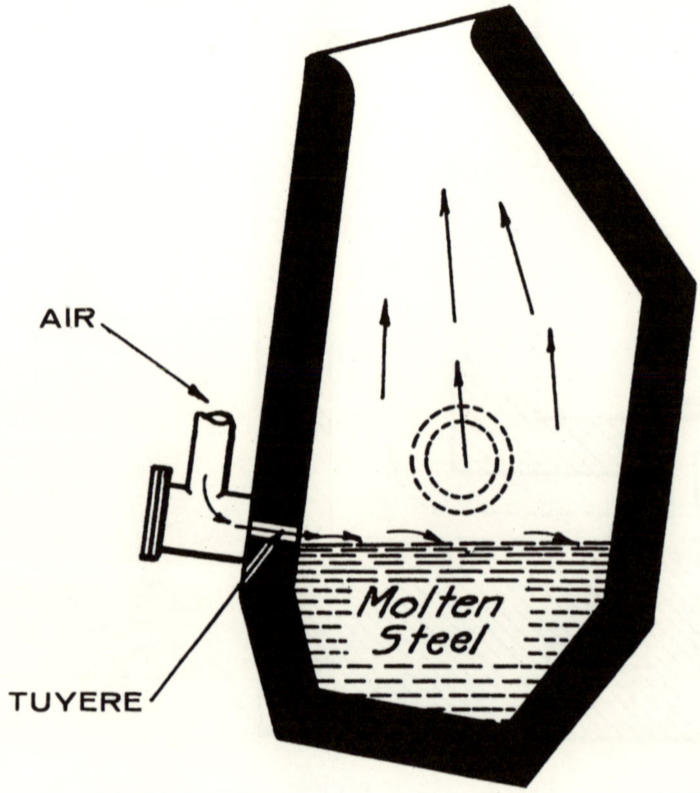

175 *Cross-section of a Tropenas converter, introduced in 1891 and used widely by American steel foundries at the turn of the present century. It featured lower pressure blast and deeper metal bath than previous converters.*

works in Eddyville, Kentucky. His theory was that iron in ore is not metallic but a chemical compound of oxygen and iron. He came to the conclusion that after the metal was melted, further fuel would not be required, but that the heat generated by the union of the oxygen in air with the carbon in the metal would be sufficient to decarburize the iron.

His friends thought Kelly crazy, and even sent him to a doctor for treatment, with the result that the doctor agreed to back him. In 1851 he was able to produce a rather soft steel, but had trouble with his carbon.

In the meantime, in England, Sir Henry Bessemer was working along almost identical lines. He, however, had metallurgical assistance. Robert Mushet of England developed an alloy of iron, carbon and manganese which purified the metal and assured the presence of enough carbon to make steel. Although Bessemer obtained American patents, Kelly proved his patent priority in 1857, and in 1866 Kelly and Bessemer joined forces.

The first use of a Kelly converter occurred in 1857-58 at the Cambria Works at Johnstown, Pennsylvania. Later Kelly obtained rights to use Mushet's patent. Bessemer's first converter in the United States was installed at Troy, New York, in 1865. This converter was introduced into foundries and was further improved by the Tropenas converter (Fig. 175) in 1891, which blew over the surface of the metal rather than through it. The lower pressure blast and deeper metal bath thus obtained gave better results.

176 *Cross-section of an open-hearth furnace fired with unregenerated gas or liquid fuel. First installed in 1870, its tremendous capacity soon surpassed the tonnages of converter steel. More cast steel is produced today by open-hearth than by any other steel melting method.*

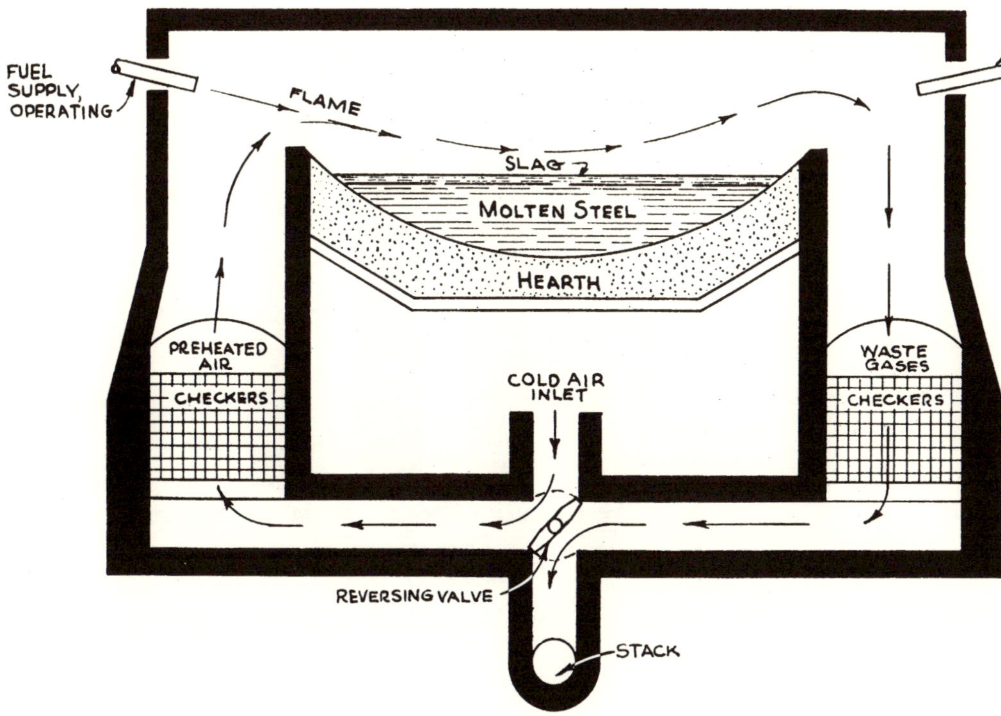

Hardly had the converter become well implanted when another furnace came into use and gave the steel industry the capacity it required. This was the Siemens-Martin open hearth, a development dating back to 1845, plus the succeeding experiments of J. M. Heath. However, it did not become successful until the great heat of the open-hearth regenerative furnace could be supplied to it. This became possible in 1857, as a result of the invention of the gas producer, patented that year.

Due to its heavy initial cost, the first open-hearth furnace was not installed in the United States until 1870 (Fig. 176). With its tremendous capacity, however, it soon surpassed the converter in tonnage, until in turn it was challenged by the electric furnace, much later.

The electric furnace also was a product of the fertile brain of Sir William Siemens, who developed an electric arc in 1878 (Fig. 177). It was little used, however, until improved by Girod, Heroult, Keller and others in 1895, and was first installed in the United States in 1906 at the plant of the Holcomb Steel Co. (now the Crucible Steel Co.) at Syracuse, New York.

177 *Cross-section of an electric-arc furnace, now extremely popular in the steel foundry industry. The first such furnace in the United States was installed in 1906 at Syracuse, New York.*

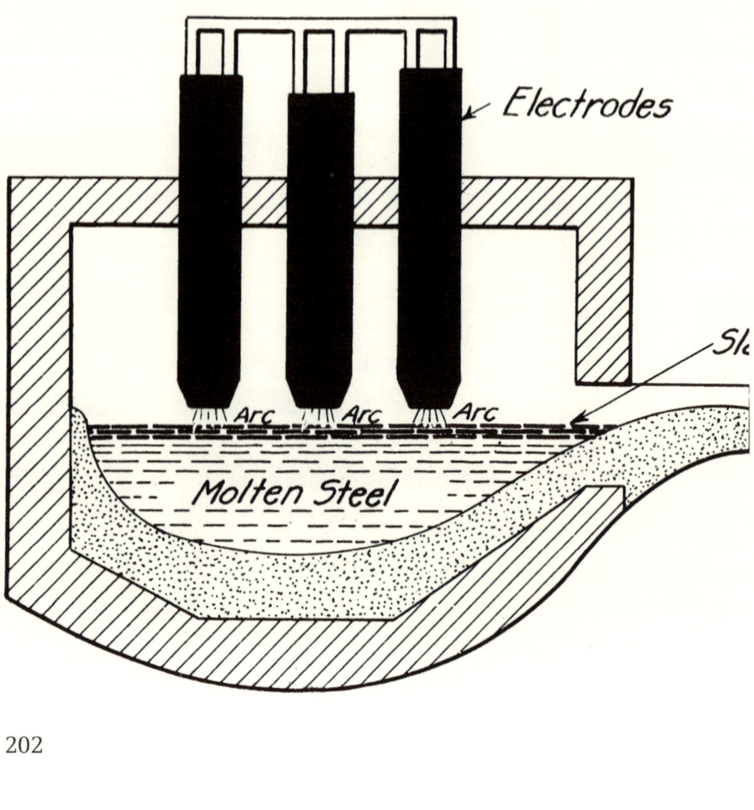

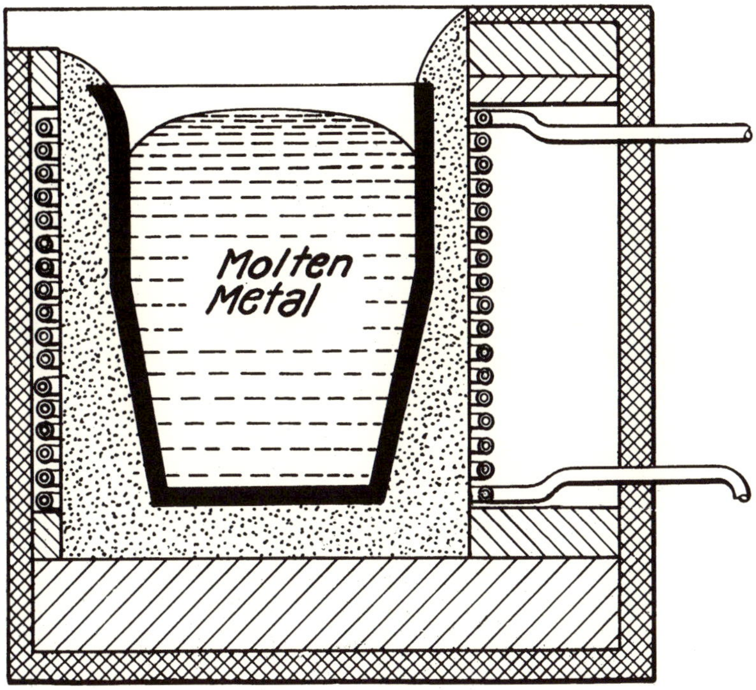

178 *Cross-section of a high-frequency induction furnace. Induction furnaces were not introduced in the United States until 1930, having been originally developed in Sweden.*

Induction electric furnaces were not introduced from Sweden until 1930 (Fig. 178), and many of the American developments on this furnace were made by Dr. E. F. Northrup and Dr. G. H. Clamer of the Ajax Metal Co., Philadelphia.

In discussing the various types of steel furnaces it is interesting to note the tonnage percentages accounted for today by the several steel-furnace melting methods:

Open Hearth	65%
Electric Arc	30%
Electric Induction	2%
Converter	2%
Crucible	1%

Steel castings came quickly into widespread use in the United States, and their production in large quantities was a major factor in the railroad expansion that spread so rapidly throughout America. In 1861 the first crucible steel was poured at the Buffalo Steel Co. (now Pratt & Letchworth Co.), Buffalo, New York. In 1867 the first commercial steel castings on a large scale, and for railroad use, were produced by the William Butcher Steel Works (now the Midvale Company), in Philadelphia. The first American foundry to produce steel castings exclusively was the Pittsburgh Steel Castings Co. (now the Pittsburgh Valve Foundry Co.), organized in 1871.

In 1888 the first manganese cast steel was made in the United States at Taylor Wharton Iron & Steel Co., High Bridge, New Jersey, under license from Robert Hadfield of England. This alloy steel was used also for railway crossings and switch frogs, and it marked not only the first cast manganese steel in the Untied States but also, so far as is known, America's first cast alloy steel.

The only other notable metal alloy of this period was a non-ferrous product, babbitt metal, consisting of 24 parts tin, 8 parts antimony and 4 parts copper. Babbitt metal was developed around 1862 by Isaac Babbitt, and was further exploited in 1877 when B. H. Dusenburg patented a machine for casting babbitted parts for machinery. At approximately 650° Fahr., the alloy was forced into a metal die by means of hand pressure on a lever. Later, to obtain better results, the babbitt formula was altered by adding zinc and increasing the proportion of tin.

Gray iron, malleable and steel having been added to the foundrymen's metals for casting to shape, it follows that equipment and methods for the rapidly growing castings industry were being given increasing attention. Molding, coremaking, sand conditioning, and metals and materials handling methods also progressed during the 1800s, and to complete the story of the development of the art and science of casting it is only logical to trace the early progress of the industry's tremendous mechanization.

CHAPTER **10**

Development of Foundry Mechanization

CONTINUOUS melting and furnace improvements over a period of about eighty years during the nineteenth century brought to foundrymen melting tools superior to any previously known. With efficient and economical melting equipment the foundry industry was able to develop a metallurgical chemistry which, coupled with the art of casting to form, made it possible to produce excellent and economical parts that proved indispensable to the machine age.

The new furnaces which were introduced during the nineteenth century did not, of course, satisfy all phases of the industry's needs, but fortunately other divisions of the foundry also were progressing rapidly. The new furnaces soon were supplemented by better blowers, pouring devices, microscopic analysis of metals, molding equipment, mechanical chargers and many other tools commonly accepted in modern foundry practice.

Metallography was first developed by Henry C. Sarby of Sheffield, England, in 1863. Chemical analysis had been available prior to this time, but Sarby was the first to polish, etch and microscopically examine metal surfaces for physical analysis. Sarby first became interested in and developed microscopy as an aid to a study of meteorites. His work on the surfaces of metals soon became much more important from an industrial point of view, since it enabled the practicing foundryman to supply the missing link of knowledge and supplement his rather sketchy experience with chemical analysis.

Metals and melting were further aided by development of blowers designed specifically to meet foundry needs. The steam engine and the water bellows had already proved of great benefit to

179 *The Roots blower, an important addition to the foundryman's mechanical equipment, put on the market shortly after the middle of the nineteenth century.*

the foundries, since they permitted the attainment of higher temperatures and shorter melting time. Mechanical blowers entered the foundry market as commercial devices some time after the middle of the nineteenth century, although home-made equipment had preceded the standardized articles for many years. These blowers were of two types: the cyclone, as typified by the Sturtevant blower, and the Roots, which is of the box-blower type. Both of these types are well known today and have been greatly improved by their present-day manufacturers (Figs. 179, 180).

Quite early in the development of pouring devices, numerous mechanical aids were invented which today are highly specialized and built with standard and interchangeable parts. The shanked ladle, adapted for use by two men, appeared later in the century; still later came the one-man ladle equipped with wheels. When it was realized, however, that foundry flooring was not ideal for smooth transportation of molten metal, foundrymen began to move their ladles overhead by crane or winch (Fig. 181).

As demands of the machine age grew for larger and larger castings, ladles of greater capacity were required, and these increased the hazard of pouring. Many accidents in foundries during the early 1800s were due to improper pouring devices (Fig. 182).

In 1867 James Nasmythe, the inventor of the steam hammer, came forward with a ladle which undoubtedly has prevented countless metal-pouring accidents. He constructed a safety foundry ladle, the tilt of which was controlled by gearing, as illustrated in Fig. 183. The foundry industry quickly adopted this device for pouring all sizes of castings, to the greater safety of men and greater economy of operation (Fig. 184).

In addition to the introduction of melting and devices complementary to metal, there occurred the important development of the molding machine, without which the modern foundry would be incapable of its present large-scale production. A detailed survey of the development of molding machines is contained in an excellent paper on *Machine Molding* by Wilfred Lewis, published by the Franklin Institute of Philadelphia in 1911.

180 *The Sturtevant blower, of the "cyclone" type, driven by a pair of vertical stationary engines. Also introduced some time after 1850.*

181 *Hand shanked ladle (above) for overhead crane pouring, and wheeled ladle (below) for one-man operation. Both devices made their appearance in the latter part of the nineteenth century.*

182 *How a crane-operated bull ladle was handled prior to 1867, when the geared-tilt ladle was developed.*

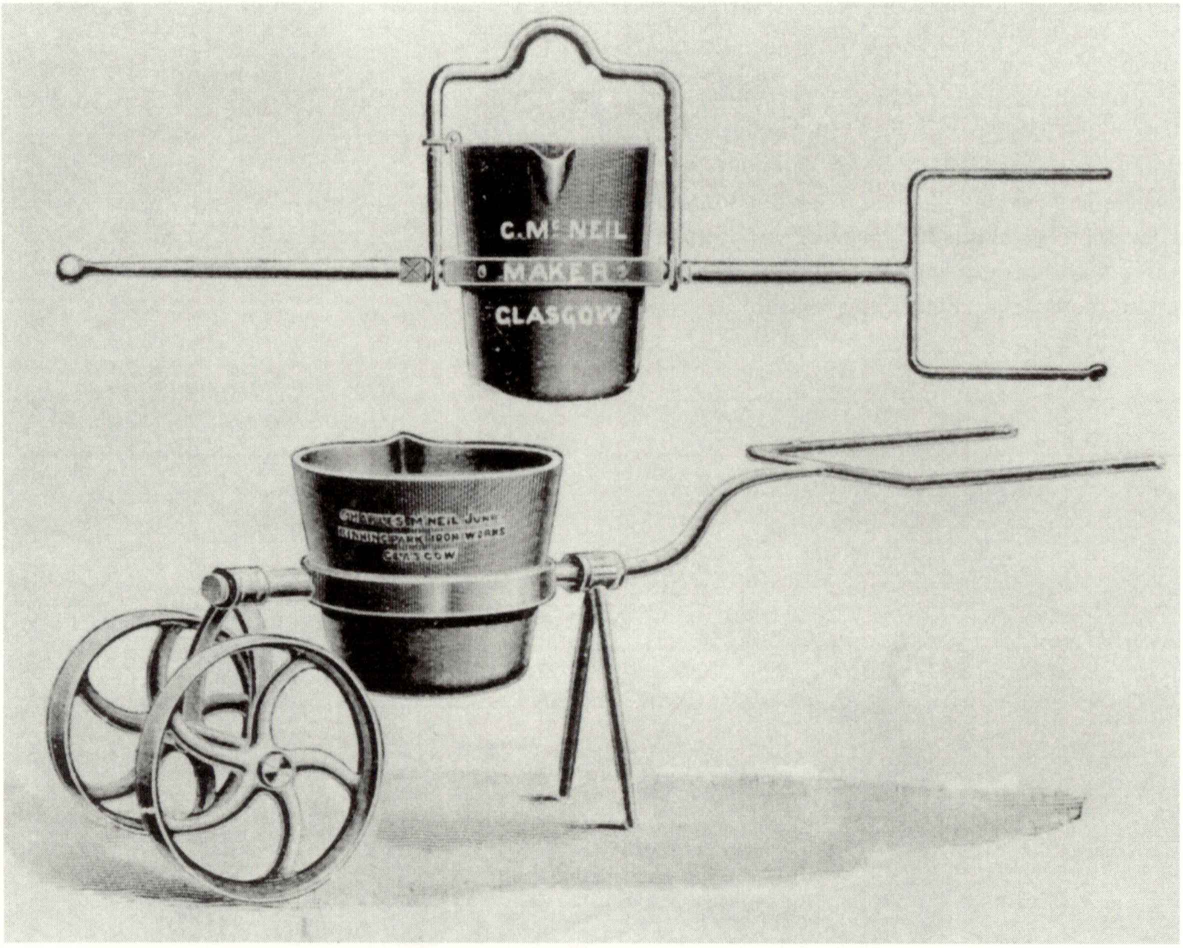

183 *Geared safety ladle, as suggested by James Nasmythe in 1867, a device which undoubtedly has prevented countless metal pouring accidents in the foundry.*

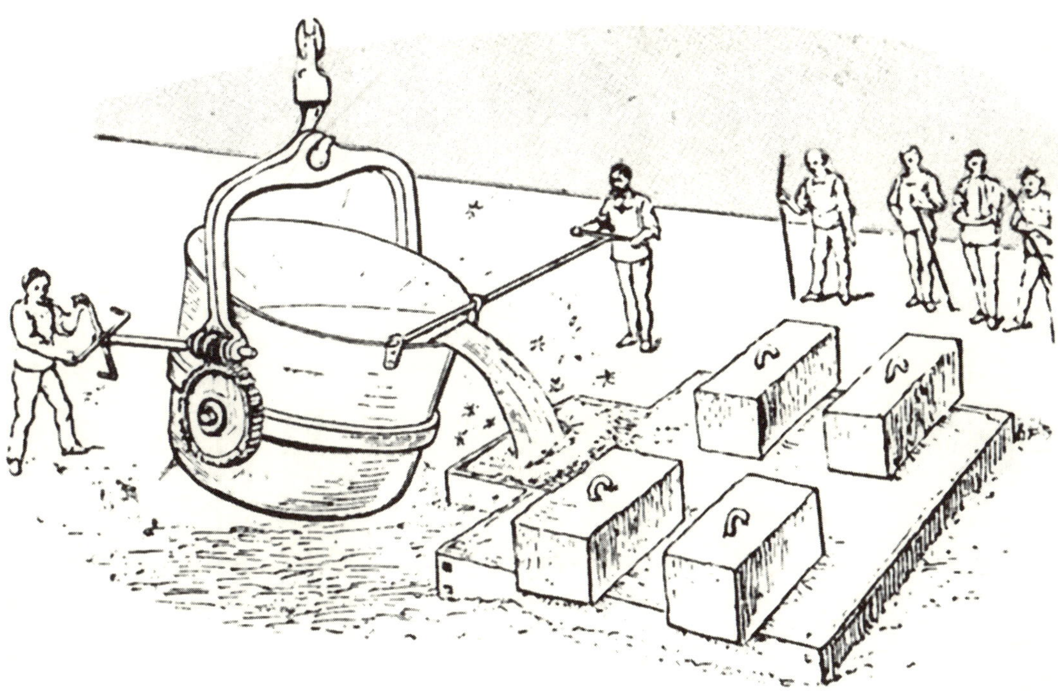

184 *Pouring steel castings from a small geared ladle during the Spanish-American War.*

Molding machines, of course, had been the foundryman's dream for centuries, but it was not until the nineteenth century that such equipment actually appeared. There is a record that an unknown Englishman developed a machine in 1800 for molding screws, a device in which the pattern was backed out of the sand by lead screws of the same pitch. Not until 1837, however, was a dependable molding machine put on the market. A jarring type, this machine first was made and used by the S. Jarvis Adams Co., forerunner of the Pittsburgh Iron & Steel Foundries Co., now the Mackintosh-Hemphill Co. While of rather crude design and built for special work, the equipment was used successfully in making a number of castings on one riser—quite a feat for that time.

In 1869 Hainsworth, another American, made several molding machine improvements, and in 1878 Adams further simplified his own machine. In the meantime, the imagination of molding machine designers was running rampant. All sorts of devices, many of which would be termed highly impractical today, were introduced. One idea involved the use of a rubber diaphragm around the mold. The principle was good but the application was poor, since the diaphragm continually broke.

185 *An early molding machine of Tabor make. From the crude molding devices attempted around 1800, the first dependable machine appeared in 1837. The practical men who developed today's highly efficient equipment did not come on the scene until the eighties.*

Finally, in the 1880s, practical men appeared who gathered all the old ideas together, added many of their own, and originated the commercial equipment common today (Figs. 185, 186).

Foremost among these men was Charles Herman, who founded the Herman Pneumatic Machine Co. (Fig. 187). Herman's picturesque (and unaltered) description of how he came to invent his molding machine gives us an excellent picture of the degree to which practical foundrymen have influenced the development of practical foundry equipment:

"At that time all patterns for bench molding was solid; sunk halfways into a plaster match bord. The molders with the pressors broke the plaster in those frames almost as fast as the could be made and instead of arranging the patterns to suit the mashines, the stored them away in some corner.

"1880 brought the stripping mashine. This mashine rigged up proper with patterns in striping plates and interchansibal flasks, doubled the output over hand molding. After this: All smart heads

186 *A Teetor molding machine, designed in 1889.*

187 *The late Charles Herman and one of the jarring machines which he invented. His picturesque description (see text) of molding machine developments illustrates how practical foundrymen have influenced the development of practical foundry equipment.*

in the foundry line sett them thinking on molding mashines. Meny mashines, and some very expensive ones were brought into the market, most of them prooving a faluare.

"As mashinist and patternmaker, and having an interest in one of the leading foundrys and mashine shops in Pittsburgh, Pa., I had the best opportunity for experimenting in this line. We had a running order for stove knobs. I made splitt patterns and put them on a steel plate, half of them on each side. I went to the foundry where I knew the had discarded six of those first pressers and bought two of them for little more than scrap price. With the patterns on steel plates and those mashines we got better knobs and half as meny more than by hand moulding.

"The first moulding mashine I made was a pressing and pattern drawing mashine, the first two combind in one. Then I made a mashine to press from top and bottom at the same time. This proved a practible mashine for deep work until the jarring mashine tood these patterns in. I did not think of making mashines for the market, but for our own patterns I wanted the best I could think of.

"One day I went trough the foundry and seen a molder, (on a certain pattern) throw the sand in the flask by his hand and got a good casting. I tought there is still another way of making molds, that could be tried on a mechanical way. I convayed the sand up to the third floor and had it coming down in a bulk the exect size of the flask.

"Under the shute stood a striping mashing with pipe balls. With this arrangement we got twice as meny pipe balls for a days work than we got before on the strip-mashine and hand ran. One drawback was a looss of 20-25 p.c. on the striping mashine hand ramed 10-15 p.c. on account of swells. The all had to be perfectly round and had to go trough a steel gage. When the pipe calls came out on the cleaning barl, a man with the gage got on top of the pile and gage them all.

"After six very busy months, I got my hands and mind free and tought of tacking it easy. I walked trough the foundry and saw a laberer wheeling the swelled pipe balls to the cupolo. I tought to myself there should be a remedy to prevent the loos. In thinking it over and over again, I came to the conclusion to harden the soft spotts in the same by piking up the mold and giving it several pumps.

"I went right to work and made patterns for a jaring and pattern drawing mashine to take in the place of the striping mashine. When this mashine was ready it was sent to the foundry to be used next morning. I was away from the works the next day and the morning after, my first tripp was to the foundry to find out how the new mashine where doing.

"The molder, an intelligent young man, always willing to do his best said to me: I did not use that mashine under the shut. A chain link on the conveyer was broke and I tought I jar up a mold and put a new link in afterwards. When I found what a nice mold I had I jared another one and got so interested that I keppt on all day. I did not get as meny molds but think very near a meny good balls.

"When the came out of the cleaning barel and the man with his gage got on top of them, he did not know what to make out of it, all went trough his gage. After this, we made all the pipe balls on jaring mashine and was convinced that there was no other method that could ram a mold as perfect. This was the latter part in 1887.

"In 1888 I aplied for a pattent and had it alowed in 1890. The compressed air was hardly known at that time, I was runing the mashine by hand leever and cams and belt power.

The later part in the 90ties brought the pneumatic chissels hamers at cc. and the air compressors into the foundry. Now was the proper time to bring the jarring moulding mashine into the market. With the air leefting power, I was not afread to build a mashine to ram up almost any pattern that was ramed up on the floor, an undertacking never tought of before.

"In december 1897 I starded a mashine shop to build the jaring mashine for the market. My friends, have seen the results I had with the J. mashine in our own foundry propezied me a clear sailing; but is was not so. As I said before: there was meny mashines, (with to much mashinery for a foundry) Brought into the market that prooved a rack inside of six months; so the foundry men where very suspisius of enything new. The called it a crazy idee to leeft pattern sand and flask and drop it continuisly, it would be only a matter of time when all had to go to pieces.

"Our foundry moulding mashine builders kept this view up more than anybody else. This sertantly acted as a brake, but the jaring mashine found it way into the foundry and proved what I calimed for. I starded in the molding mashine business with my own invention. I starded where the presser stopt. We never buildet a molding mashine of enybody else make."

Charles Herman typifies the kind of man who made modern foundry practice possible. We must also recognize the men who worked with him and who, between 1890 and 1905, brought out the molding machines that now are universal names wherever foundrymen meet. They are, of course, Tabor, Pridmore, Teetor, Mumford, Beardsley and Piper. Their designs and persistent efforts to improve molding practice have lightened the foundry task and enabled foundrymen to increase production, produce more accurately and uniformly on a production basis, and to lower costs. Without such hard-headed practical inventors the modern foundry could not meet present-day demands.

Developments in molding logically include the use of permanent molds, although actually the permanent mold preceded the loam and sand mold by centuries. Subsequent types of permanent molds gradually made their appearance, but for many years they were limited in application by the metal available. Permanent molding may be defined simply as the pouring of a liquid metal into a metallic mold, usually preheated, and permanent-mold methods include centrifugal, slush, die and Cothias casting.

The pioneer of the centrifugal casting process, which involves the pouring of molten metal into a rapidly rotating metallic mold, was A. G. Eckhardt of Soho, England, in 1809. The method was early adopted by the pipe foundries, being used first in Baltimore, Maryland, in 1848. Sir Henry Bessemer, famed for his converter, used centrifugal casting to remove gases and was the first to pour two or more metals into a single rotating mold.

Centrifugal casting of steel was first attempted in 1898 at the plant of the American Steel Foundries in St. Louis, Missouri. Car wheels were spun-cast in 1901 by a Mr. Needham of Rockford, Illinois, who advocated a speed of 620 revolutions per minute as the proper rate.

Following the early development of the centrifugal method, two further permanent-mold methods were introduced. One is known as slush casting, which involves the pouring of liquid metal into a metallic mold and, after a skin set, inverting the mold and pouring out the unsolidified center mass of metal. The method seems to have been in use long before the nineteenth century.

A still later development was Cothias casting, of limited application, which provided for a metallic mold and a permanent core which was forced by a plunger into the metal while still in a liquid state. Thus has developed the permanent-mold or die-casting procedure, an important auxiliary to foundry operations, which also involves the forcing of molten metal into metallic molds.

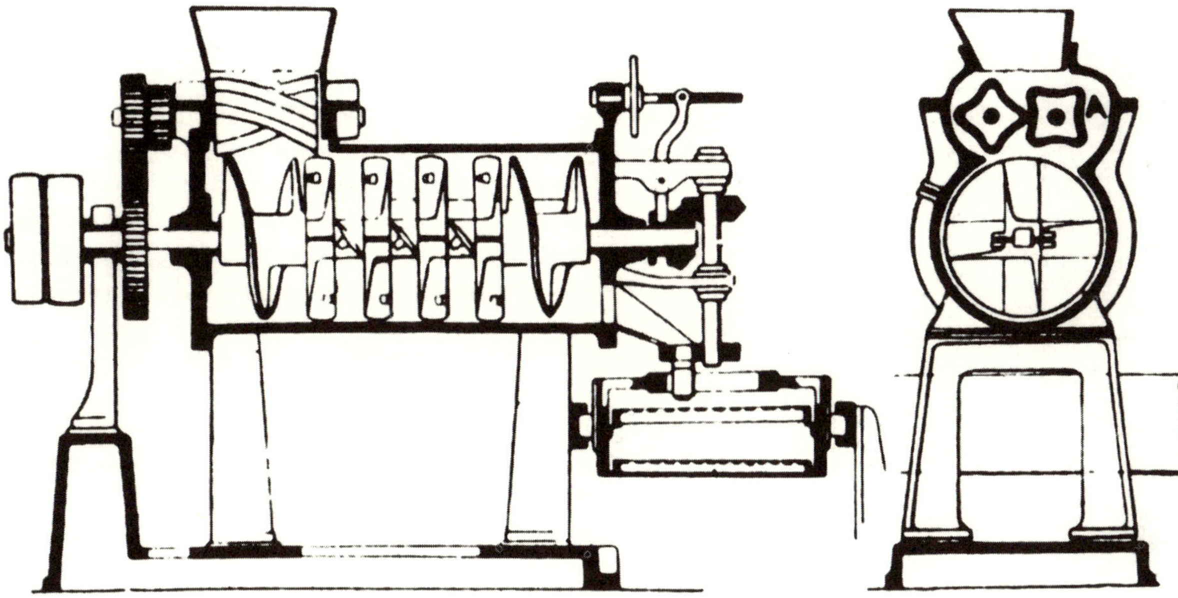

188 *Early machine for grinding, mixing and riddling coal and sand for facings. Such paddle mixers did not appear until the seventies.*

Competitive methods of casting, melting and molding can and do exist side by side, not for the purpose of eliminating any one process but rather with the broader aim of widening the overall scope and importance of the entire foundry industry. Consider, for example, the various types of sand now available as molding materials. From time to time the manufacturers of sand preparing and handling equipment are likely to bemoan the use of permanent molds, or plaster molds, or cement molds. This attitude seems no more well founded than does the competitive point of view that gray iron, brass, bronze, steel or malleable are, in all ways, highly competitive as engineering materials. Each has a place in industry adapted to its particular virtues and cost. In the fact that all foundries together can offer a great variety of cast materials, made by many processes and with a remarkable range of properties, lies the great and underlying strength of the industry.

Sand castings are relatively recent (in terms of the foundry's great antiquity) and occupy an indispensable place in the industry. As we have seen, at an early date foundrymen undoubtedly realized that a mixed molding material was essential, and during the

Middle Ages wool clippings and then manure were added and beaten into the loam with a bar. Under melting temperatures the combustible material burned out, leaving a porous mold. The word "permeability" was unknown then, but today we realize that those early founders were actually striving to improve the permeability of their sands.

Under the influence of loam the early machines for the treatment of sands aimed at compounding or grinding rather than mixing and mulling. Early in the 1870s, machines began to appear which were essentially paddle mixers, an example being shown in Fig. 188. These gave way to a centrifugal type of machine in which iron balls were spun around to grind the clay, brick and coke used with the sand in those days. Fig. 189 shows one such mill.

Following this development came the centrifugal aerator made by Sellers of Philadelphia, but here the manufacturer went to the opposite extreme and no real combination of sand and bond was effected (Fig. 190). The aerator merely brought the materials into close proximity, without achieving a new or compound product.

In the 1890s and early 1900s manufacturers began to utilize the edge runner, adapted from the ceramic industry, both with and

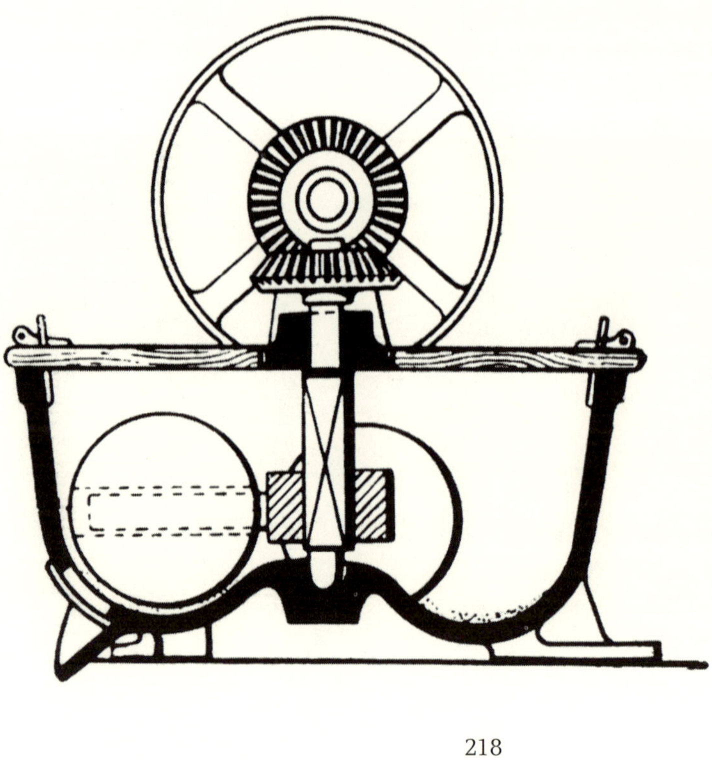

189 *Evans centrifugal type mill in which balls were used to pulverize brick and clay for sand additions.*

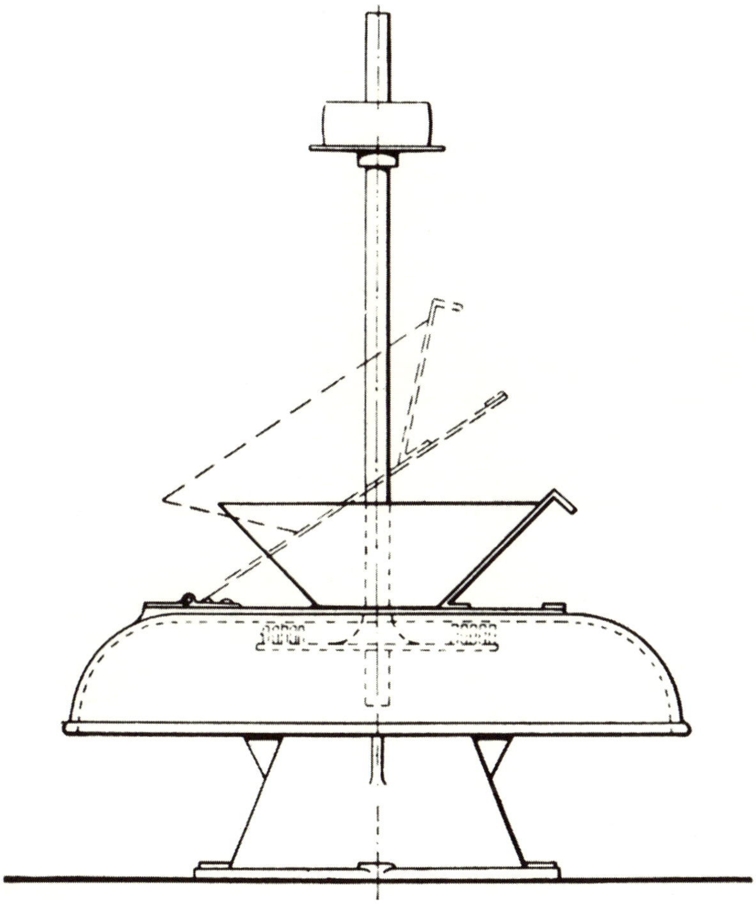

190 *Centrifugal type aerator-mixer, made by Sellers, which effected no real combination of the sand and bond.*

191 *Smedley Bros. revolving pan mill with a web-foot muller, introduced around 1900. The first muller employing individually mounted revolving mullers made its appearance in 1912.*

192 *Early belt-driven sand riddle, developed about the beginning of the twentieth century.*

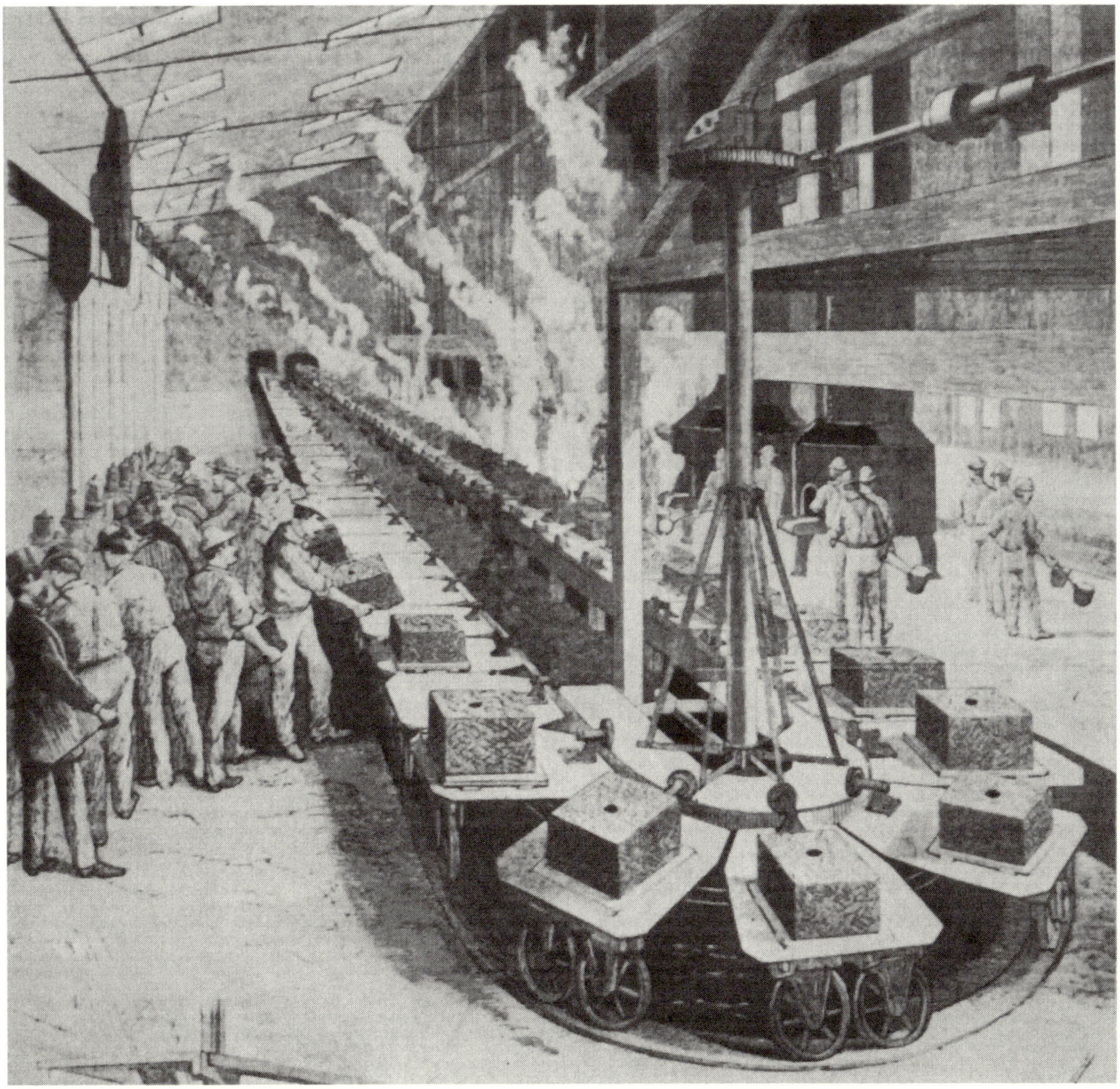

193 *Mold conveyor of the type first installed in 1890, showing integrated molding, pouring and cooling operations.*

194 *Shakeout station on a mold conveyor of a half century ago. Note hexagonal screen and sand elevator in the background for returning sand to the molders.*

without a revolving pan (Fig. 191). In 1912, however, the first muller with individually mounted revolving mullers of varying weights was put on the market by Peter L. Simpson. Since that time mullers which effectively coat the sand grains have been used successfully in the preparation of sand for both cores and molds.

This same period saw the first steps taken toward development of screening machinery, which eventually resulted in the riddle, the magnetic separator and the complete sand-preparation plant (Fig. 192).

Mold-conveying methods were introduced about 1890, originally involving a continuous series of moving cars which looped at a steady speed from molder to pouring station, then to cool and shake-out. American assembly-line methods, of course, have greatly simplified and improved this procedure (Figs. 193, 194).

Core manufacture had also been standardized and equipment made available for mass production in American shops. Long rooms filled with boys producing cores in mass could be seen as early as 1888, a typical core room of that date being shown in Fig. 195. All cores were racked and dried in kilns and then placed on racks to be carried to the foundry.

The importance of core production soon gave impetus to the development of specialized baking ovens particularly adapted for

195 *Mass production of cores by boys, a common coreroom sight in 1888. Cores were made in hand molds, dried in kilns at the rear of the coreroom, and racked as at left in picture.*

196 *Millett core oven, invented by Eli Millett in 1887, for drying small cores in individual drawers. Today, coremaking and core drying are highly specialized divisions of the castings industry.*

197 *Sly tumbling mill, considered the first such cleaning mill in the foundry field, developed between 1880 and 1887 by W. W. Sly of Cleveland, Ohio. Cleaning equipment enabled the foundries to reduce greatly their hand chipping and grinding operations and to produce a finished product.*

foundry use, so that cores of different types could be baked longer or shorter as required. An early developer of the foundry core oven (Fig. 196) was Eli Millett in 1887. Today, coremaking, like sand molding and conveying, is a complex and highly specialized division of the foundry.

It seems certain that early credit for the tumbling mill must be given to the W. W. Sly Co. of Cleveland, Ohio (Fig. 197). The Sly cleaning machine was a great boon to foundrymen, since it enabled them to offer their customers a finished product. The importance of this development can be better appreciated when the modern foundryman recalls that no electric torches were then available and all grinding and chipping was accomplished by hand.

In addition to tumbling mills for small castings, the sandblast was developed for larger work by R. E. Tilghman of Philadelphia in 1870. The use of such equipment did not begin to broaden, however, until about 1900, when it was installed at the Logan Manufacturing Co. at Phoenixville, Pennsylvania (Fig. 198). Of the rapid improvements made since then by equipment firms, mention should be made of the American Foundry Equipment Co. (now the American Wheelabrator & Equipment Corp.) of Mishawaka, Indiana, and the Pangborn Corporation of Hagerstown, Maryland, whose recent innovations have added much to the ease and efficiency with which foundries now handle many cleaning room operations.

198 *Tumbling department of a large foundry in 1888. The sandblast, which first appeared in 1870, was not used extensively until about 1900.*

199 *Foundrymen—a very special breed.*

CHAPTER 11

Automation and Aerospace -New Image of Metal Casting

THE TWENTIETH CENTURY began without portent of the dramatic changes computers and automation would bring about by the 1960s. Alterations of equipment and methods themselves would be quite obvious—the change in foundrymen more subtle. In education, pay and social standing by 1968 foundrymen were far removed from the illiterate laborers who pressed, desperate for employment, into the American foundries during the great immigrations of the 1890s and early 1900s.

As man began his journey among the stars (in space vehicles that use metal castings) there were still many alive who could recall foundrymen slaving in smoke filled foundry caverns—sweaty, soot-smeared faces glistening in the furnaces red glare, battling in personal combat the very elements for their existence. Men like these in the metalcasting town of Pittsburgh as early as 1858 founded the Sons of Vulcan trade union. Among hard, picturesque men they were a very special breed.

This remembrance of yesteryear, man bringing forth unaided new images into the world, like Prometheus from the foundry's fiery maw, rivals the most heroic sagas of all ages. Unfortunately, along with the romantic concept go all the drawbacks animal man is heir to.

In the year 1900, with the life expectancy of American males at age 49 (by 1968, 66.5), the accident rate of foundry workers was more than triple what it is today. Those giants strained mightily 60 hours a week for an annual take home of $500. Should they feel exploited they had but to glance about at the 6.5 million in a population of 76 million who had no income at all or recall that America's first metal casters worked for 15 cents an hour, sunup to sundown, 6 days a week.

In 1968 few foundrymen in America would be denied work. Having enjoyed a 40-hour week since 1932, with automation and work simplification well upon them, there were less than 3.8 million unemployed out of 200 million. A foundryman's position was never so secure; never had he been so important or so well protected.

As the twentieth century began, the average US foundry poured more tonnage than was cast throughout the world when the Nation was born. From 30,000 tons* in 1775 to 50,000 tons in 1800; to 1,135,956 in 1864 and to 13,789,242 in 1900.

By the mid fifties the average foundryman's takehome pay, proportionately, exceeded corporation earnings. By 1954 it was eight times that of 1900; the economy was sound and purchasing power was two to three times over the 1900 figure.

In 1963 gray iron casters in the United States were making $2.78 an hour for an annual average individual wage of $5,782—a little over ten times that of 1900 and nearly twice the national average income for that year. In that same year $2,017,016 worth of gray iron castings emanated from 1140 foundries employing 120,430, who earned altogether $730,084,000. Steel castings—$861,022,000 worth—were shipped from 266 establishments employing 47,564 whose total pay was $360,591,000. Copper-alloy foundries, 549 in all, employed 15,341 workers with a payroll of $86,350,000 and shipped $287,534,000 worth of castings. Malleable iron, meanwhile, cast in 83 establishments, shipped $312 million worth of castings and employed 22,651 who would that year be paid over $144 million.

Despite so huge a transformation man himself was called upon to expend far less energy. For a minimal effort he produced in less time more and more sophisticated shapes for more and more intricate machines.

As automation took over production rates climbed, until one automated foundry in the automotive field in 1967 was able to establish a consistent production figure of six-manhours-per-ton of castings. Such modern foundries carry no molders on the payrolls, instead they are operators of molding machines, that themselves perform the thankless toil.

Metal machines replaced the labor of man and horse and with the sudden impetus given metal manufacturing in World War I machinery became a necessity. Without cast metal parts, machinery (the machine age, for that matter) could never have existed.

First came machinery, followed by mechanization and automation, and with automation came a dramatic increase in output and

*pig ingot must be used since no reliable figures exist for casting themselves this early. It is safe to presume, however, that at least one half went into such cast objects as cannon, bells, stoves, kitchen pots, ships anchors and sundry parts. Data for this comparison is from *History of the Manufacturing of Iron in All Ages,* by James Swank, former secretary of the American Iron and Steel Association.

wages. Not only did automation emancipate the foundry worker from dulling labor the Industry discovered it could train a machine operator in weeks as compared to the years of apprenticeship needed to turn out a skilled molder.

Metal casting took naturally to automation and did so rapidly. Characteristically, the first fully automatic plant in the United States (one of the first in the world) was a Rockford, Illinois, foundry that cast hand grenade hulks for the army in 1918. The inseparable relationship between metal casting, machinery and automation can be discerned in the progress from S. Jarvis Adams Company's 1837 molding machine to that of Hainsworth in 1869 to Adams again in 1878, to those depicted in Fig. 185-187, and on to today's fully automatic molding machines, Fig. 200.

The history of metal casting discloses the foundryman as eager as any manufacturer to survey and take full advantage of inventions; even many times inspire them. We will recall that America's first commercial metal caster, Joseph Jenks (p 164, 166), was awarded the first patent in America.

Thomas Edison called Seth Boyden "one of America's greatest inventors" for Boyden established two basic industries in America—that of patent leather and malleable iron. Among his numerous inventions, aside from blackheart or American malleable iron, was one for the first steam engine cutoff governor, a grate still used for the manufacture of zinc oxide, an inexpensive method of manufacturing "Russian" sheet iron, a hat-forming machine and a goldlike copper-zinc (or tin) alloy known as oroide.

In 1851 James Bogardus's factory in Chicago constructed with cast iron supports opened the way for what many art historians consider to be America's only original contribution to the arts of the world—the skyscraper. Bogardus also invented a pyrometer, a dynamometer and the dry gas meter in addition to cast iron supports for floors and window frames.

Organizationwise, the twentieth century would witness the demise of small family-affair foundries and the swift growth of merged, joined, absorbed and otherwise consolidated complexes which, though considerably fewer, produced vastly greater quantities of quality castings.

By 1960 less than one percent of the foundries in operation were a century old. The trend continued as huge conglomerates entered the picture, merged companies, divisions and subdivisions of unrelated production interests whose powerful combined assets made the industry of the United States a worldwide politico-economic force to be reckoned with.

200 *Automated CO_2 foundry with mold shooter (left rear), core shooter (right rear), gassing, core setting and mold closing (center rear), automated pouring (left front) and shaking (right front).*

201 *A fully automated molding machine capable of turning out more than 100 molds an hour.*

In three years, 1960-1963, though copper-alloy foundries dropped in number from 580 to 549, the survivors grew larger and more powerful, as they produced more than the total of the previous foundries. Typical of such growth was one Ohio malleable iron metalcasting plant that sold $9 million worth of castings in 1939, $31 million in 1962 and according to a projection made by its president in 1967 by 1970 would be selling $134 million worth.

Low profit-taking would make much more of this possible. A survey made in 1968 by Modern Casting magazine (official publication of the American Foundrymen's Society) disclosed that the leading US foundries had that year returned $454,000 each to plant and system modernization.

Big business and the acquisition of huge personal fortunes reached a peak shortly after 1900. The New Carnegie Steel Company, incorporated in 1900 in New Jersey, had assets of $160 million, at the time a vast sum. One year later, U.S. Steel Corporation was capitalized at $1.4 billion; for many years the largest holding company in the United States. Until the advent of personal income taxes, despite the Sherman antitrust laws, big money stayed in the hands of a few.

Metal, specifically steel, cast and fabricated, for the first time in the history of the world became, in the United States by 1900, the basic industry. By World War I, America produced more manufactured goods than the three closest competitors, the United Kingdom, France and Germany, combined. Simultaneously, along with the annexation of Hawaii and Puerto Rico and the exploitation of seemingly inexhaustible deposits of high-grade Lake Superior ore, the United States found itself for the first time a major world power.

American metal casting was big business. After a walking tour Walk Whitman described the Nation: "colossal foundry, flaming fires, melted metal, pounding trip hammers, surging crowds of workmen shifting from point to point, waste and extravagance of material, mighty castings; such is a symbol of America."

In the first year of the new century foundries in the United States poured more openhearth steel than did those all over the United Kingdom—almost as much as the rest of the world combined. To the 23.6 million tons of United States steel, Germany poured 14.7, the United Kingdom 6.4 and Russia and France 3.8 each. This ascension has been unbending ever since (barring temporary slide-backs during economic depressions) and in 1968

American metal casting grossed well over $9 billion. Employing some 250,000, metalcasting was soon to be America's sixth largest industry.

As far back as 1864 the military foundry at Old Fort Pitt (Pittsburgh) had cast a 20-inch, smoothbore Rodman cannon weighing 115,000 pounds. This was a hundred times bigger than the famed Urban Gun of Mohomet II that felled the walls of Constantinople in 1453. Three years later the Krupp plant in Essen, Germany, poured a 50-ton cast steel cannon and the fate of the French Army in the War of 1870 was sealed.

Fortunately, however, the biggest consumer of metal castings was not the military but the internal combustion engine which, in the automobile industry, in ten years gave a greater boost to metal casting than had cannon, bells and the steam engine in a century and a half. About 25 percent of all metal castings produced in this century have been component parts of automobiles, trucks and tractors.

After a somewhat slow beginning—in 1900, there were 5,000 automobiles, with less than ten miles of concrete roadway in the entire Nation—at the same time 20 million horses and 10 million bicycles—by 1948 the Industry had provided vital castings for over 100 million passenger cars.

In 1924 Henry Ford made a million autos in 132 working days. His first million took seven years. Metal casting knowhow and the world's first mass production were vital to this phenomenal production increase. Automobile output in the first ten years increased 3,500 percent, with a corresponding demand for castings.

On the rear bumpers of automobiles came trucks and tractors and other mechanized farm and industrial equipment; followed in rapid succession by parts for such mushrooming industries as refrigeration (1930s), aviation (1940s), air-conditioning (1950s) and in the 1960s data processing, electronics and aerospace technology. Cast metal played, as it continues to play, a very vital role in each.

Though the age of electricity began in 1890 by 1900 there were still many steam and hand-operated cranes. Not until the 1930s did electricity and internal combustion engines such as gasoline and diesel replace steam in almost every application of power.

Expansion and growth of the industrial work following the worldwide upheaval of 1914-1918 required better tools and metal parts. Castings had to be made stronger per pound, bigger, thinner-walled, cheaper and they had to be delivered faster. This took a casting expertise that developed apace with the need.

202 *Cumbersome electric cranes of the early 1900s took up floor space.*

Metallurgy began to come into its own in 1889 when nickel was alloyed to make a stronger steel, as family secrets, foremen's fancies, handicraft and just plain mumbo-jumbo were gradually superseded by casting technology based on controlled experimentation. The empirical ways of finding out how to do things the hard way was, by 1920, already old-fashioned to those metalcasting establishments that were to survive.

Everyone recognizes metallurgy today as the basis of sound metal-casting technology but in the beginning it was welcomed only by the more advanced; as one Simpson Bolland, who in 1894 read a paper at a meeting of the Western Foundrymen's Association in Chicago containing this exhortation:

> There can be no question as to how some of our foremost firms are producing castings for such low figures at this time. Analysis of the materials employed is at the bottom of it. By reason of the knowledge thus obtained they have been able to purchase both pig iron and scrap at ridiculously low figures, very much of which under the old rule-of-thumb methods would have been rejected as unfit for the purpose.

Bolland urged his fellow foundrymen to consider the relative importance of metallurgy. "It is not supposed," he conjectured, "that the average small foundry will employ a chemist [metallurgist] . . . [however] proprietors' sons who have hitherto been satisfied with the regular routine of office work will now have their ordinary course of education supplemented by a course in metallurgical chemistry and thus qualify themselves for the position of chemist as well as clerk."

Beginning with the use of nickel (1889), of chromium to make shock-resistant armor steel (1890) and with stainless steel (Henry Brearly in 1912) metallurgy was to introduce to metal casting a whole series of versatile alloys.

Industrial metallurgy was quite an awakening to the average foundryman; still there would be for time to come little that had not been done, at least in principle, by his grandfather. Even the computer existed in primitive form prior to 1900. For the next half century the foundry scene would be altered slowly at first and then magically, it would seem, by adoption and development of fairly old precepts.

A foundry trade magazine in 1918 ran the bold headline, "What proper routing does for an old foundry . . . System is the appeal to the efficiency man, as vital in the foundry as in the clearing house." That same issue advertised airy, well-lighted foundries—skylights to let light in and smoke out.

The old has a way of not giving in suddenly to the new and the World War I concept of a clean, well-lighted place was overlapped all to often by such diehards as that depicted in the center of Fig. 203. This foundry along the Hudson was pouring as late as 1928.

During the birth of awareness of the value of metalcasting technology, there were formed a number of Industry-sponsored organizations dedicated to metal science and R&D which, its leading members realized, could be turned to commercial advantage.

The American Foundrymen's Association (since 1948 *American Foundrymen's Society*) was formed in 1896 out of the Foundrymen's Association of Philadelphia, itself but three years old. The New England Foundrymen's Association was formed that same year and one year later the American Malleable Casting Association (changed 30 years later to the Malleable Iron Research Institute and in 1934 to the *Malleable Founders' Society*). In 1900 the Carnegie Research Scholarships of the Iron and Steel Institute were founded, followed by the *Steel Founders' Society* in 1902, the *Foundry Equipment Manufacturers' Association* in 1918, the Gray Iron Institute in 1928 (changed in seven years to the Gray Iron Founders' Society and in 1964 to the *Gray and Ductile Iron Founders'*

203 *Top to bottom, entrance to a foundry of the late 1960s, a country foundry operating as late as 1928 and an excerpt from a 1918 advertisement showing how ideal environment was aspired to that early.*

Society), the *American Die Casting Institute* in 1929, the *Alloy Casting Institute* in 1940, the *Non-Ferrous Founders' Society* in 1943, *Foundry Educational Foundation* and *National Castings Council* in 1947, the *Investment Casting Institute* in 1953, the *Society of Die Casting Engineers* in 1954 and the *Ductile Iron Society* in 1959.

In the 1967 AFS Hoyt Memorial Lecture it was predicted that automatic pouring would become more and more common and the casting of such metals as titanium would soon be on a sound production basis. Substitute permanent molding and die casting for automatic pouring and the new light metals, aluminum and magnesium, for titanium and you have what would have been a logical prediction in 1900.

Aluminum, most abundant metal in the earth's crust, was a development of this century. Isolated in 1825, it derives its name from the Latin alumen, bitterness, the same root as ale and alum. Aluminum was first exhibited in 1855 but for many years was so difficult to obtain it was more costly than gold. In 1881, for instance, it sold in the United States for $8 a pound and Napoleon III ordered the lightweight exotic metal to replace silver in his table setting. As a commercial metal dependent upon electricity for its production, aluminum could not have been made practical before the invention of the dynamo.

In 1888 the Pittsburgh Reduction Company, offering the metal in half-ton lots for $2 a pound, got few takers until one manufacturer discovered it made good, inexpensive tea kettles. Within five years the price was down to 62 cents a pound and by 1900 to 32 cents.

In 1890 only 62,000 pounds of aluminum were produced in the United States. The production was low until World War II but by 1963 $635,934,000 worth of aluminum castings were shipped in the United States. This industry, undreamed of in 1900, in 1963 employed 35,970 in 951 plants, with a payroll of $221,567,000. In the first seven months of 1968 alone 908,235,000 pounds of aluminum were cast in the United States, and this does not take into consideration the many other ways it was formed or fabricated.

Although the Wright brothers in 1903 had a 152-pound cast aluminum crankcase in the first airplane ever successfully flown, few foundries had the interest or knowhow, nor were they equipped to cast such now familiar shapes as aluminum air-cooled cylinder heads, automobile wheels and engine blocks. Hard pressed by the sudden demands of the war and blessed by almost unlimited R&D

funds, Government owned foundries led the way, inspiring ingot vendors to come up with such highly castable alloys as A1-Si-Mg.

The development of magnesium as a casting metal parallels the history of aluminum. During World War I it sold in the United States for $5 a pound and by 1935 only 375,000 pounds had been cast. By 1944, however, the industry was turning out over 86 million pounds a year, a good portion of which was cast.

Castings Shipped in the US (tons)

(Showing a trend to lightweight metals)

	1950	**1960**	**1967**
Steel	1,459,382	1,392,385	1,857,480
Gray Iron	12,905,562	11,593,511	14,327,843
Copper-Base	508,090	381,054	483,224
Malleable	920,512	820,701	1,039,903
Aluminum	226,305	375,901	767,357
Zinc-Base	268,199	311,601	418,808
Magnesium	6,157	11,770	19,990

Manually operated casting machines were patented as early as 1849 (Sturgiss) and 1852 (Barr) in efforts to satisfy the insatiable demands of a growing reading public by way of rapidly-cast lead type. These early inventions led to Ottmar Mergenthaler's Linotype, an automatic casting machine in which molten lead is forced by piston stroke into a metal mold.

The first diecasting machine by that name was patented in 1905 by H. H. Doehler (on permanent display in the Smithsonian Institution, Washington D.C.). Two years later came E. B. Wagner's casting machine, prototype of the now familiar hot-chamber diecasting machine. It was first used on a large scale during World War I for binocular and gas mask parts.

Zinc alloys were used for die casting as early as 1907 but were not competitive until Price and Anderson developed the Zamak diecasting alloy in 1929. Production then got under way fast. At this writing, approximately 70 percent of all aluminum cast and 93 percent of all zinc is produced in diecasting machines.

Investment casting (*cire perdue* or lost-wax), one of the oldest casting techniques, was rediscovered in 1897 by B. F. Philbrook of Iowa who used it to cast dental inlays. Industries paid little heed to this "arty" process until the urgent military demands of World War I overtaxed the machine tool industry. Then shortcuts were needed to provide finished tools and precision parts, avoiding time-consuming machining, welding and assembly.

204 *An automatic diecasting machine (1968).*

Interest slacked off following the emergency until, in the 1930s, the idea of making duplicate dispensable patterns was developed. When the great intricacy demanded by sculpture and dentistry was found to be highly reproducible, industrial precision investment casting was born. Investment casters were soon supplying the delicate castings needed in aviation and aerospace industries to the tune of some $300 million worth a year.

The twentieth century saw the refinement of processes and materials used in the foundry for over 400 years. Biringuccio (pp 86-90) recommended the dregs of beer vats and human urine as binders for molding sand and both were in use well into the twentieth century. It is curious to note that both are, in modified forms, still used. Since breweries were unable to keep up with the demands for dregs, such substitute organic compounds as furfuryl were developed. Ureaformaldehyde (replacement for urine) is somewhat more agreeable and is now supplied to foundries by the large chemical houses.

Until the 1920s sand testing consisted of feeling—squeezing a handful to judge its ability to compact and stick together. Early in that period a sand research committee of the American Foundrymen's Society began to develop sand test methods (see "*The Era of Sand Testing,*" H. W. Dietert, AFS TRANSACTIONS, p 637, v 75, 1967). By 1924 a set of standards was established for the Industry that covered moisture content, strength, permeability and fineness. Sand aerators and throwers came into general use after Peter Simpson's sand muller (p 222), followed by the invention in their Oregon works in 1912 by E. O. Beardsley and W. F. Piper of the Sand Slinger.

Other important developments and innovations of the first half of this century were continuous melting, better and larger mold and core blowers, spectography (pioneered for metal analysis early in the 1930s by professors at the University of Michigan) and mechanical charging of cupolas.

Some time earlier pulverized coal, fed in by powerful blowers, had been installed as a fuel, first at the Iron and Steel Manufacturing Company, Lebanon, Pennsylvania. Ovens replaced open fires for drying cores and internal heating plants took the place of the time-honored custom of heating with open-air, unvented salamanders. Chilled iron shot and hydroblast began to take the place of sand blasting, with its latent threat of silicosis. Band saws were being used and compressed-air-powered chipping tools along with belt-driven hand-operated grinders (still less efficient than the 17,000 plus sfpm of the late 1960s).

Electric furnace production rose from 27,000 long tons in 1914 to 511,695 in 1918, out of a total steel ingots and castings of 23,513,030 (1914) and 45,060,607 (1917).

Sir William B. Siemens is credited with being the first to use the electric arc for melting metal in a closed hearth. The first American electric furnaces had flat back walls that all too often caved in when they were tilted for tapping. Electrodes were square without joints and the stubs had first to be entirely withdrawn before new ones could be inserted, thus interrupting melting. On April 5, 1906 the first heat was tapped from an electric arc furnace in this country (p 202) and by 1909 there were 77 installed. Though at first arc furnaces were used exclusively for expensive carbon tool steel their use soon spread (by 1917, 139 in the United States) and before long the crucible steel process became obsolete.

The first low frequency electric furnace installed in the United States was at the Crucible Steel Casting Company's Lansdown, Pennsylvania, plant in 1913 for special melting. The first high-frequency coreless electric induction furnace was installed in 1930 at the Lebanon Steel Foundry, Lebanon, Pennsylvania.

Today atomic power promises still more efficient electric melting. Perhaps it was more than coincidence that in 1954 the first atomic power plant to be built anywhere in the world was at the Dusquesne Power Company is Pittsburgh, since that was the scene of some of the Nation's earliest casting successes.

Since the turn of the century the United States has been at war five times. Each conflict was destructive of material and human beings alike but though each disrupted a smooth trend of peacetime production each gave impetus to invention. Regrettable as wars are, throughout history they have been the spawning ground of metallurgical innovation. Cannon founding and its peacetime alternate, bell founding, gave foundrymen in the olden days what expertise they possessed. Krupp developed cast steel in order to pour superior cannon (1847), even as Henry Bessemer was spurred on by a standing money reward offered by Napoleon III for steel to match that of the German warlords.

Electric arc furnace tipped for pouring.

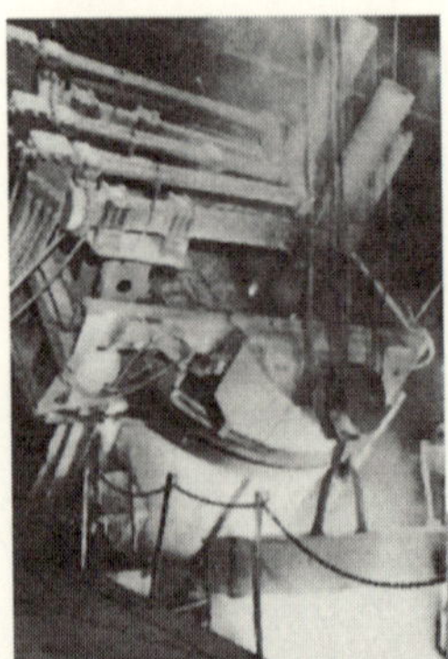

205 *World's largest (1967) high vacuum induction furnace used for superalloys.*

The science of metal casting has been used by wartime concentrations of money and brainpower to further national effort in each conflict. Each time the legacy of the emergency period has been a peacetime bonanza. The U. S. Government financed five-sixths of new plant construction during World War II and by war's end owned 90 percent of the facilities for aircraft, synthetic rubber, magnesium and shipbuilding, 70 percent of the aluminum capacity and half the machine tool facilities. This was largely given away at prices as low as 20 percent or written off to the civilian operators. Steel ingots and castings reached the peak output in 1944 of 80,037,130 long tons, almost 50 percent above 1929 output, but the most striking industrial gains of the war occurred in alloys and light metals. Aluminum production, for example, increased sixfold, magnesium from 2,500 tons in 1939 to 170,000 tons in 1943.

Wars, expanding production facilities and economy unnaturally, were followed by more or less severe economic depressions. One, the so-called Silver Campaign Depression, was slowly being overcome as the new century began. Few, however, rivaled that which began with the collapse of the New York Stock Exchange in 1929.

No basic industry suffered so during the troubled 1930s as metal casting. Coinciding with the money pinch, the 1930s also saw the emergence of competitive materials; plastics and the wholesale use of weldments. Indicative of the plight of the Industry were the steel foundries whose output fell off a quarter million tons.

Hundreds of metalcasting establishments cooled their fires permanently in the Great Depression, as the market for castings became smaller and smaller. Foundries merged to stay alive, in fact if not in name; or they consolidated or were taken into receivership and in order to survive at all learned to diversify. Whole plants became captives of other industries, most notably automobile manufacturing.

During the first 40 years of the twentieth century guesswork, casting "art" and pure unadulterated mystique gave way to science and its direct application, technology, as muscles deferred to brains and to mechanical and electrical power. Without this emphasis, metal casters could scarcely have made the response to such increasing metallurgical and casting demands as are represented in the progression from wagon wheel hubs to steam engine journal boxes to automobile wheel hubs and on to such sundry space vehicle castings as yaw bellcranks.

206 *Experimental foundry in the General Motors Research Laboratories Metallurgy building.*

Stiff competition during the Great Depression along with the upcoming second worldwide conflict emphasized the importance of research and development. R&D became the magic word that would open the doors to progress and economic well-being. The total effect of the R&D, applied this early, is graphically portrayed in Fig. 211.

In 1968, 90 percent of all the scientists who ever lived were busily employed on one project or another (or they were teaching others). In the decade 1958-1968 more inventions would be conceived than in the prior history of mankind. Research was no longer the self-willed efforts of a scattered few and one-man brain shops like that of Thomas Edison had long since blossomed into elaborate independent departments of multimillion dollar corporations.

The $125,000 plus allocated by the American Foundrymen's Society in 1968, though it was the largest single amount expended for metal casting research, was but a drop in the bucket compared to the millions upon millions spent by private organizations on applied research. This in turn was a pittance beside the billions spent on research directly and indirectly by the United Sates Government, in 1940 Government and industry combined were proud to have spent $280 million on assorted R&D. Twenty years later federal agencies alone spent close to $3 billion, and in 1968 $8 billion; planning by 1970 to be putting out over $20 billion a year.

Instruments and testing procedures replaced local ground rules and hunches. When the information machine, the computer, was finally marketed the data it made accessible caused less and less duplication of metalcasting research the world over, as had occurred, for instance, in the independent discovery by Bessemer and Kelly of the converter. Coordinating bodies were set up on a national scale, the *National Science Foundation, National Defense Research Committee* and others, all of which began with the *War Industries Board* of World War I and the *National Academy of Sciences.* The *War Production Board* through the World War II *War Metallurgical Committee* sponsored such notable work as that of C.O. Burgess and T. E. Barlow, of the Steel Founders' Society. New data published by these two caused the *American Society for Testing and Materials* to raise the upper limit of gray cast irons from a prewar 60,000 psi to 100,000 psi.

Hope was given for worldwide cooperation in metal casting by the formulation in 1923 of the *International Committee of Foundry Technical Associations* (Zurich, Switzerland), which strives through 24 nations and an annual *International Foundry Congress* to exchange technical data.

The stone ages lasted thousands of years and the metal ages more thousands of years, but the ages of discovery today are reckoned in less than decades. We are in the passage of a number of ages that go on simultaneously—atomic age, electronic age, jet age, computer age and the age of aerospace research and travel.

The speed and nature of today's travel place extreme demands on metal casting. Greater thermal, chemical and mechanical stresses are met almost faster than they are planned for, as castings are now not only airborne but moonward bound.

During World War II the inoculation of gray iron became common practice, as high quality cast irons replaced the more scarce steel in many castings. Shortly after the war a new type iron, called variously spheroidal graphite cast iron, nodular iron and now more universally ductile iron, was patented and announced by International Nickel Company. It was a major breakthrough in metallurgy for its high strength and ductility allow it to compete with malleable iron and even in certain applications with steel. By 1967, 863,000 tons were produced in the United States and another 811,000 in Japan.

Metalcasting research is goaded by the active interest of the military services in what are initially purely military applications—turbo-supercharger blades, miniaturization and air-transportability

cutaway view of a Project Mercury space capsule, showing just a few of the essential metal castings.

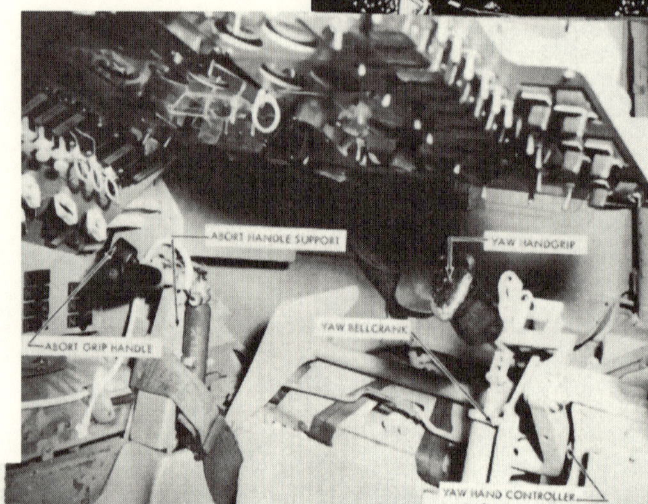

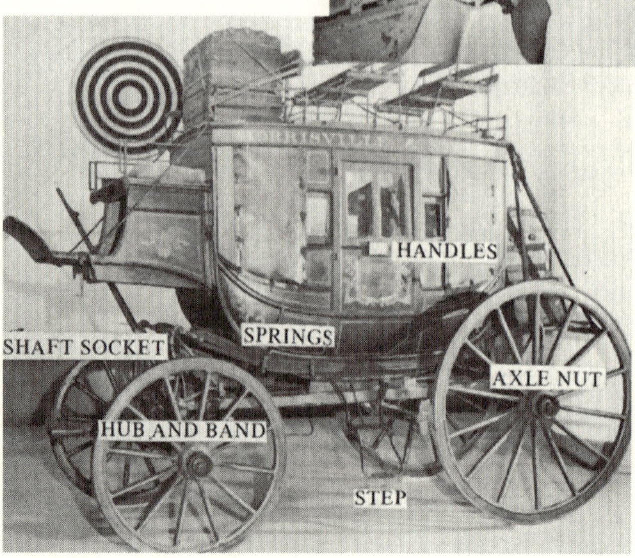

207 *From stagecoach to staged space capsules. Man's long journey depends upon metal castings.*

of equipment and machinery. One of the first large-scale uses of aluminum was canteens for American servicemen in the Spanish American War.

New, sturdy casting alloys have evolved using tungsten, cobalt, nickel, chromium and molybdenum and since World War II experimentation has been accelerated in organic and chemical sand binders for thermosetting. Cold air-setting cores and molds evolved. Phenolics led the way to urea and the dielectric process followed by furan and urea-free resins. Many patterns were made of epoxy resins and polyurethane and other expendables as well as such full mold materials as polystyrene. The latter came about in 1958 after H. F. Stroyer was granted a patent for the use of polystyrene for patterns rammed up in the mold. Upon contact with hot metal they melt and vaporize leaving no trace.

208 *This foamed polystyrene pattern weighs 45 pounds, the outcome will be an 8-ton iron casting.*

In 1966 alone, over 170 million pounds of new type binders were used in the hot core processes, including core oils. Sixty million pounds of such binders as the furans were being used in cold box, hot box, nobake and CO_2 processes.

Interest grew in stack-molding and ferrous die casting. As controlled directional solidification became a reality, exothermic materials were used around risers to provide better insulation, thus reducing riser size and increasing metal yield. Thin wall cylinder sleeves were then feasible for aircraft engines. At the same time jet engine turbine blades, for the hot end, were being cast in single crystal metal.

In the 1950s arc cutting torches all but replaced pneumatic chisels in the cleaning of steel castings. Later on EDM (electric discharge machining) was employed with success for pattern and die making.

Magnetic analysis was used as early as 1915 to detect flaws in iron and steel castings. It would soon thereafter be adopted by the U.S. Bureau of Standards, as would the application of x-rays, made practically one year later by William V. Coolidge's workable tube. The million-volt x-ray tube was developed during World War II and on-site use of gamma ray and radium capsules. Radioscopes, such as cobalt 60, by-product of atomic energy research, has became common in metal casting for radiographic analyses.

As this is being written General Electric Company scientists are perfecting a nondestructive testing device name neutrography that employs a stream of neutrons, subatomic particles, from an atomic reactor to enable an inside picture to be seen many times more sensitive than x-rays. Held back at present by their immobility, when atomic melting becomes a reality, neutrographic devices will go hand in glove as superior testing means.

Today the Industry is forced to reassess the time-honored definition of metal casting—act of pouring molten metal into a mold. New technologies of metal forming, some of which border on casting, present a serious challenge that shakes whatever sense of complacency may linger.

General Atomic, subsidiary of General Dynamics Corporation, has developed high energy magnets that form metal not by conventional pouring but by jolting with magnetic pressure. Named Magneform, it is the first entirely heatless casting method and will enable metal to be cast on to rubber, plastics and even explosives. A similar metalforming method is one that makes use of superplasticity. Through carefully controlled heat treatment metal becomes plastic enough to be drawn into metal dies by the negative pressure of vacuum.

New and sophisticated welding processes have proven a great adjunct to metalcasting. Laser (light *a*mplification by *s*timulated *e*missions of *r*adiation) with millions of degrees of temperature presents immense possibilities. Though it can vaporize diamonds with ease, the extreme heat aspect of laser is not as remarkable as its controllability; so great it is used as a cutting instrument in the most delicate of surgeries.

Along with the laser cam electronic welding. Inertia or diffusion welding, a method developed by Caterpillar Tractor Company, uses spinning to combine metal parts by the heat of friction so tightly x-ray cannot detect the union.

Zircon, chromite and olivine foundry sands followed the introduction of bentonite into the American Castings market.

During World War II steel foundries learned to substitute castings for forgings. The process was continued in peacetime. Rigid ordnance requirements became a leading factor in the broadening use of x-ray, ultrasonic and other nondestructive testing methods. After the war the U. S. Department of Commerce helped introduce to the United States the so-called C molding process, named after its World War II German inventor, Johannes Croning—now referred to as shell molding and coremaking.

The demand for high strength-to-weight ratios of military ordnance parts was carried over into the exacting aerospace engineering field, and the almost new technique of water-quenching steel came into general use. Heat treating itself was a most significant innovation both in methods and equipment, for with it properties could be maximized with minimum dependence on alloying. Martempering (quenching in lead or salt) was introduced into many ferrous foundries, as both batch type and continuous conveyor lines were installed to handle the rush of work. Rigorous government inspection requiring micro-tolerances brought about a whole new concept of quality control along with stepped-up, inplant R&D.

As soon as foundrymen got into the habit of applying science in its many facets to metal casting they steadily learned to gain control of their work. Predictions became commonplace and quality control departments sprang into being along with time-and-motion study. Guarantee of top quality became so prevalent many foundries offered to reimburse customers for unplanned for machining costs. Such par excellence is today typified in the Zero Defects programs of many alert metal casters.

209 *Using a mobile ultrasonic reflectoscope on large castings in the yard of a West Coast foundry.*

210 *Cobalt gamma rays for radiographs of castings. This equipment can handle castings up to 8 inches thick.*

About the time aluminum was being discovered the Englishman Charges Babbage, builder of the Analytical Machine, the world's first computer, observed: "the errors which arise from the absence of facts are far more numerous and durable than those which result from unsound reasoning respecting true data." As steam heralded the first industrial revolution so the computer was most prominent in the information explosion that catapulted the second industrial revolution. The results were as dramatic as the LD type basic oxygen furnaces that, complety computer controlled, make a normal heat run within an hour.

By the late 1950s it was obvious there had begun a second, more fantastic industrial revolution—machines making, repairing and running themselves, controlled by elaborate, almost living it seemed, electronic brains. The most dangerous and obnoxious jobs were relegated to automated on-the-spot robots, programmed like human workers to lift, carry and pour. Cupolas could now be charged and discharged not only mechanically but automatically.

Beginning in the late 1950s, as this tireless information machine helped reduce the brute labor force, gross national product climbed rapidly, in proportion to a renewed technological effort. This did not reduce employment totals, as labor leaders mistakenly predicted, but created new, more interesting, higher-paying jobs. It elevated the stature and the very nature of the foundryman's job.

211 *The thrust of change.*

courtesy, Gray and Ductile Iron News.

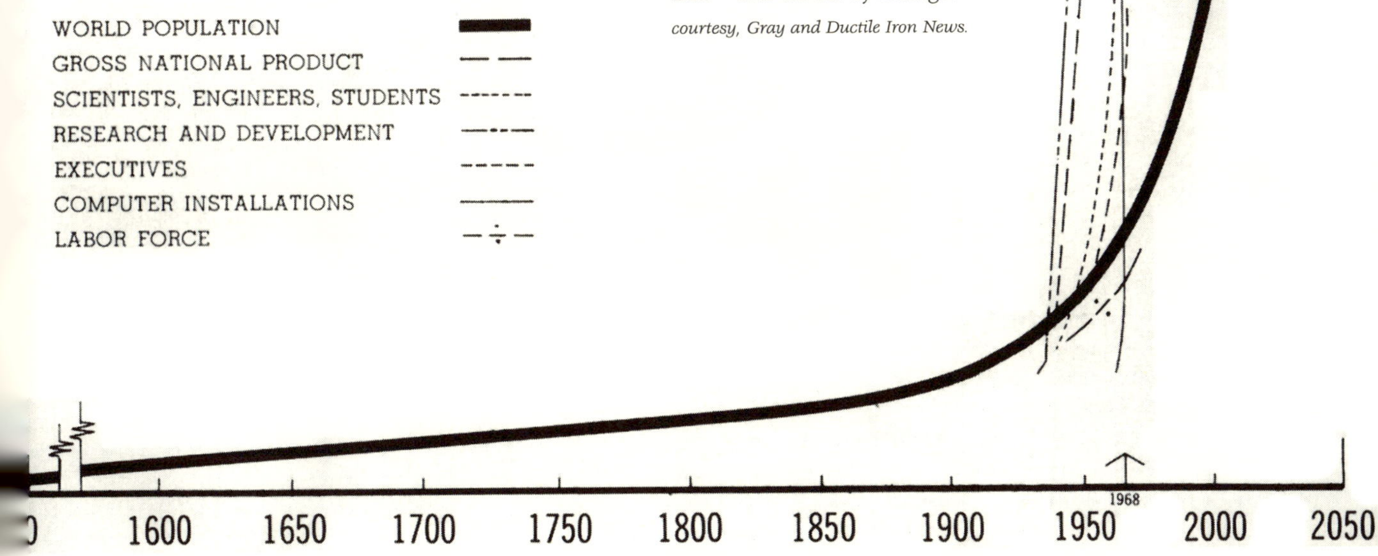

212 *Computers (bottom) generate and store the information vital to programming the casting process, while realtime information is fed in constantly from the foundry floor to keep the data current. Operator (left) uses pushbutton panel to monitor an otherwise automatic process, while two men in the middle picture study plans under a flow chart of an automated cupola operation.*

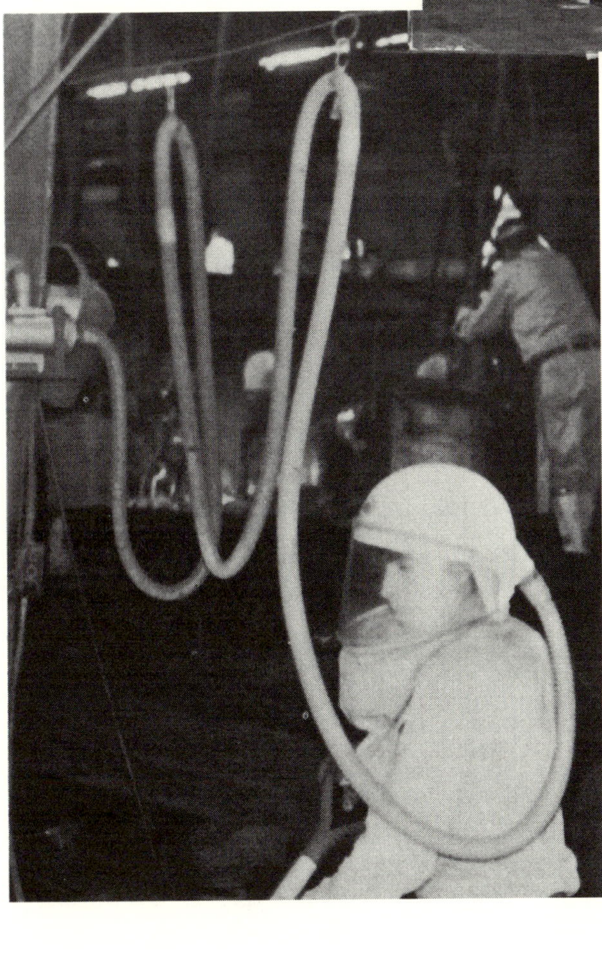

213 *Air pollution controls inside and out. Schematic photograph shows essential flow-path in an air pollution control unit. The man in the space suit, below, is a shakeout operator who breathes clean air and manages to keep cool in a hot, dusty atmosphere (which, too, is carefully filtered before leaving the plant).*

214 *Cast aluminum, thin-wall automobile engine block being removed from permanent mold.*

By this time no single man could hope to cope with all that mankind knew. No man could hope ever again to catch up to the aggregate human mind or to the total memory of the metalcasting industry; nor in fact could man unaided hope to produce enough to fill his ever-growing needs. Computers were the only answer. By 1958 there were 1,000 of these electronic brains in use in the United States and 160 in all of Europe. Ten years later there were 35,000 installed in the United States with a prediction that by 1975 there would be 170,000.

Preheated metal, customized scrap and the delivery of molten metal via truck or train became standard procedure and through such electronic and electro-mechanical peripherals of the computer as servo motors many foundry operations became online, real-time, continuously automatic, from ingested raw materials at the back door to loaded finished work on the dock.

To eliminate potential Bottlenecks computers were programed to analyze the placement and configuration of gates and risers and predict and prevent expensive breakdowns. Along with the changed emphasis from muscle to mind, as the world grew smaller and smaller came the quickened interest of foundries in community relations. Metal casters, whose efforts once were proudly proclaimed by the size and density of plumes emitting from their smokestacks became public-relation-conscious and concerned over their civic responsibilities to help abate the growing pollution of water and air. To further this critical end, the *American Foundrymen's Society* retains on its staff experts in water and air pollution to assist foundries in meeting and solving environmental control problems.

In the 1960s came high-pressure molding, fast-setting nobake binders for sand, cold box process, permanent steel molds, hot box core and mold making and a standardization of casting processes that permitted high predictability. Casting improvements with the new more permeable sands permitted thin-wall automobile engine blocks and cast aluminum engines.

There has been more casting progress since World War II than in the previous 3,000 years. Still the demand goes on for foundries to surpass themselves in quantity and quality. Metal casting remains a basically essential industry as necessary to space vehicles as it was to mankind's very first machines. As science is applied more and more, so the skill, education and pay of foundrymen go up and ever up, increasing in turn the total wealth of the world.

Certainly, metal casters have much to be proud of. But as Benjamin Franklin warned the scientists of his day, "the world is now increasing in experimental knowledge but let no man flatter the age with pretending we have arrived at the perfection of discoveries."

* * *

The main purpose of this history has been to tell in condensed form the fascinating story of metal casting. It should assist today's foundryman to gain a deeper appreciation of the field in which he operates, so foundry employees and their families, as well as customers, will better understand the improvements made in foundry technology.

No one studying the metalcasting industry can fail to be impressed by the tremendous scope and importance of its products, by the tenacity and inventiveness of its workers and, most important of all by its unsung position. If the Industry now assumes in the reader's mind its proper stature—a very basic, economically essential, timetested position—the purpose of this book will have been accomplished.

Perhaps the vast importance of the metalcasting industry would better be emphasized if foundrymen and the public alike were to attempt to imagine a world without cast metal. Such a world as it now is would not exist. It is the responsibility of today's foundrymen and those affiliated with the metalcasting industry to preserve and advance that technology for the benefit of mankind.

Source Material

A select bibliography of private papers and other primary sources, periodicals, institutional publications, books and pamphlets used by the author are obtainable by writing to the technical publication editor, AFS. A majority of these can be found in the stacks of the American Foundrymen's Society metal casting library at the Des Plaines headquarters.

At this date it is impossible to ascertain the whereabouts of the owners of all the original illustrations used in this work. Many, in fact, have died since the first edition. Not all the original illustrations are retained in such formerly credited sources as the annual Transactions of the American Foundrymen's Society or the Penton Publishing Company. Some quoted sources are now defunct, as the British Foundrymen's Association and Canada's Foundry Journal, and a number of the books acknowledged in the 1948 edition are too long out of print to be accessible.

Individuals interested in tracking down any of the illustrations herein should contact the technical publications editor, American Foundrymen's Society, where either the originals or reproducible copies of all illustrations are maintained. Following is a partial list of known sources—

source	**illustration number**
American Iron and Steel Institute, NYC	2, 17, 137, 141, 142, 145-147, 160-162, 164, 165
Bettmann Archives, NYC	4, 8, 37-40, 43, 52, 53, 76, 85-87 ,136, 152, 182-184, 193-195, 198
Garland, Bannister, Griffin, Ltd, London *Ancient Egyptian Metallurgy*	45-47, 62
Lamprecht Collection, American Cast Iron Pipe Company, Birmingham, Alabama	127-130
McGraw-Hill, NYC, *Man and Metals* (1932)	5, 6, 42, 44
Metals and Alloys (1940)	3
Museum of Natural History, Chicago	9-11, 26-31, 36, 48, 49, 54-61, 148, 149
Pirotechnia (English translation 1942) AIME, NYC	64, 65, 70-72, 77-83, 95, 96
Professional Photographers of America, Inc.	199
Technical Publishing Co., Ltd., Manchester, England, *Modern Iron Foundry Practice*	179-181, 188-191
University Press, Oxford, England, *Gunfounders of England*	7, 90-94, 98, 99

Index

a

b

b c

c

d

e

f

f

g

h

i

i j

j

k

l

m

m n

n

o

p

r

s

t

u

v

w x z

w

x

z